SOMETHING NEW UNDER THE SUN
Building Connecticut's First Solar Home

SOMETHING NEW UNDER THE SUN
Building Connecticut's First Solar Home

RUTH F. FRANK

BRICK HOUSE PUBLISHING COMPANY, INC.
Andover, Massachusetts

Published by Brick House Publishing Company, Inc.
3 Main Street
Andover, Massachusetts 01810

Production credits
Editor: Shirley Keddie
Copy editor: Sandra Hughes Minkkinen
Book design: Herb Caswell
Cover design: Ned Williams
Cover photo: Robert Perron
Typesetting: Together Graphics, Inc.

Printed in the United States of America

Library of Congress Cataloging in Publication Data

Frank, Ruth F. 1915-
Something new under the sun.

Bibliography: p.
1. Solar houses—Connecticut. I. Title.
TH7414.F7 728.6′4 79-22222
ISBN 0-931790-03-4

Contents

To C.B., my husband, for his foresight and steadfast support; to Harriet F. Pilpel, my sister, for her wise and generous counsel; to Everett Barber and Donald Watson for successfully piloting us on our solar course; and to my children and grandchildren who hold the hopes and promise of our solar future.

R.F.F.

Introduction

I wrote this book to tell the world that solar heating works! When the possibility of solar heating our "dream house" was first mentioned to us, I admit I was skeptical. But I soon became convinced that solar is wonderful. And I've written this book to tell you all about it. Not only has our solar home proved to be an excellent financial investment, it has brought immeasurable gains to the quality of our lives through our everyday closeness with nature.

Since we first started planning our solar-heated home more than six years ago, solar building has mushroomed. And we are convinced that many thousands of solar homes will be built within the next few decades. I wrote this book for all potential solar builders, today and in the years to come. I want to share with you what I learned about solar building: how to choose a site, how to position the house in order to take greatest advantage of the natural elements, how to integrate solar features into the design and layout, and how to find the right solar builder.

I shall try to answer for you (in nontechnical language) many of the questions we had when we first walked out onto the solar stage: What exactly is solar radiation? Where do you find it? How do you collect it? Can it be stored? And can it really save the homeowner a lot of money?

We have learned all this and much more since we embarked on our solar venture. But my husband and I learned the hard way: by groping and grumbling, faltering and stumbling, while often being too ignorant and uninformed for our own good. I'd like to help other solar home builders avoid our many pitfalls and problems which we experienced largely because we were pioneers in this field.

The book is more than just a statement or even a strong vote for building a solar house. It is also a book about building any house. The reader can learn how house plans and specifications are drawn up, how to set a realistic budget and stick to it, what "change orders" are and how to avoid them, the ins and outs of building contracts, and how to go about furnishing and decorating a house.

For solar consumers, the book will also tell how and where to find a solar architect (if you need and want one), how to put together and work with a solar team, and what you should include in a solar heating contract.

For all those who do decide to go solar and for all those who may be considering it or just want to know more about solar building, we want to share our experience. We want everyone to know how building and living in Sol-A-Sea, one of New England's first solar homes, has saved us money, enriched our lives, and awakened us to the wonderful world of solar energy.

We hope you, too, will let the sunshine in!

RUTH F. FRANK

1

Embarking on Our Solar Adventure

Living in a solar home has brought us a fresh and comforting awareness of our dependence upon nature and our basic ties to its rhythms and changes. We've come to feel a new kinship with our natural surroundings. As seen through our wide expanses of glass, the beauties of the shore keep us closely attuned to the cycles and patterns of nature—the ebb and flow of the tides, the phases of the moon, the changing of the seasons, the sunrises and sunsets, and the migration of the birds.

Over the past five years we have become great admirers of this natural extravaganza, sometimes trading an hour's sleep for the joy of watching the sun come up. At first only the palest rays and subtlest hints of a reflected glow appear above the water. Soon these dissolve, only to reappear as bright and piercing shafts of light, and finally the whole bursting brilliance of the sun reveals itself.

If we are lazy enough to remain in bed late, we can look up and watch the clouds and birds sail by our high clerestory windows in the brightening northern skies. The sun is our faithful companion from the moment we wake up to sunlight streaming through our east-facing windows, until late afternoon when the sun begins to descend and soon disappears into the sea. We are always conscious of its cheerful, warming presence.

My husband, C.B., and I first began thinking about building a beach house almost ten years ago—although at the time a *solar* beach house was the furthest thing from our minds. We were both looking forward to retirement within the next few years. Our married daughter, Marianne, had her own family, and our son, Steve, was finishing prep school and looking toward college. Our ten-room colonial home in Stamford, Connecticut, seemed less appealing and it had always failed to meet one important criterion for us: it was not on the water.

The shore, boating, and swimming have been recurring themes in our life together. We met on a sailboat, honeymooned on a freighter, and have spent several vacations on cruise ships. C.B. has owned a boat almost continuously since he was eight years old, and I have had a passion for swimming since I was five.

When we began thinking about a retirement home on the water, we envisioned a modern but not too unusual house with cathedral ceilings, large brick fireplace, and a modest price tag. What we got was none of the above, but rather an open, airy house with light flooding virtually every room and the stars and clouds in clear view through our ceiling-high windows—at a price much higher than we'd bargained for and yet more of a bargain than we'd ever dreamed.

Had we known the aggravation, delays, and unplanned expenses that would accompany building one of the first private solar homes in the Northeast, we might not have been so anxious to build. But then again, had we known we could sit in our living room with an outdoor temperature of 30° F. and an indoor temperature of 70° F. without using a drop of oil for thirty-six hours, I think we might have wanted to do it sooner! But of course that would have been impossible. The state of the solar art in the late sixties and early seventies meant that we were pioneers, adventurers, and, in part, guinea pigs.

When we actually embarked on our building adventure, it was mid-September, 1969. At the time we were not sure if we wanted to build or buy a home. We began looking for a waterfront home near Pilots Point, on the Connecticut shore halfway between New Haven and New London. We had spent several vacations at the Point and liked the area. Shorefront houses are always hard to come by since they are often passed down from one generation to the next. When they do come on the market, they are usually so overpriced or antiquated that they are poor risks. While we could not find a single home on the water for sale, we did discover two empty seaside lots. C.B. encouraged me to research the records at the town hall and find out the names of their owners. I did this and we later called one of them. He informed us that he had been searching up and down the Atlantic coastline for years and now that he'd found just what he wanted he had no intention of relinquishing it. The owner of the other lot turned out to be a doctor from another state. His nurse returned our call and informed us that he wasn't interested in selling either.

After a couple of weeks had passed, C.B. urged me to call the doctor again. He might have changed his mind. This time his wife answered. I must have called at precisely the right psychological moment. It was eleven o'clock in the morning and her husband, an obstetrician, had not yet arrived home from the hospital where he had been all night. She listened to what I had to say and answered promptly, "Of course we'll sell the lot. We'll never get down there anyway. He's still at the hospital. He's been up all night!" She assured us that he would call us back later that day.

The call didn't come until early evening. The doctor was brusque and businesslike. "I understand my wife said we'd sell you our lot on Pilots Point. I guess she's right; we never seem to get down there. The price is twenty thousand dollars."

That was a shocker! We knew that land values were inflated, but we also knew that some homes on the Point had been selling for little more than $20,000 just a few years earlier. I asked if the price was negotiable. "Negotiable! I've been paying taxes on that property for five years and I intend to make a good profit on it. Besides, I didn't call you. I don't care if we sell it or not; you're the one who wants it so badly. Take it or leave it." After a quick, muffled conference, we told the doctor we'd take it.

Handling the usual legal and financial details took many weeks. Finally, on December 24, 1969, we became the proud owners of our precious piece of land on Pilots Point—a fitting Christmas tribute to each other.

Now that we owned the land, it no longer seemed so urgent for us to build. In the two and a half years that followed, we carried on a running romance with our lovely little beach site. Despite its limitations in size—it is only 60 by 90 feet—the vast expanse of sand and sea gives the magnificent illusion of open, endless space. Spectacular views abound in a 180-degree arc.

We learned to love our seaside land and soon got acquainted with it in its many different moods. In the winter with a dusting of snow it was serene and desolate; in early spring we could bask in the brilliant sunshine and drink in the fresh sea air. We had picnics on the beach in summer and swam in the cool, clear water. Sometimes at dusk we'd stay to watch the last purple glow of the sunset only to turn and find the dazzling reflections of the full moon as it slowly rose behind us.

By the summer of 1972, we'd come no closer to building. We were, however, becoming aware of some danger signs in the world around us. The country's ever-growing appetite for energy was making headlines. We were increasingly dependent on foreign oil; our own fossil fuels—oil, coal, and gas—were becoming scarce and were rapidly increasing in price. But somehow no one seemed very concerned. It was still beyond the comprehension of most people that our big, rich, powerful country could be courting disaster. However, the Arabs and other oil producers had already formed OPEC, a strong coalition of oil-exporting countries. Like many others, we too found it impossible to believe that a handful of less developed countries could threaten not only our standard of living but our very way of life.

With the national scene in such a state, it hardly seemed the right time to build. But C.B. is an optimist, and in spite of the gathering storm clouds, he felt we should get on with our building project. From the beginning, almost everything seemed to go wrong. First, without having any detailed idea of the kind of house we'd like, I set off to look for a builder with only a rough sketch in hand. The sketch showed the barest outlines of a beach house, a two-story affair with a steeply pitched roof. The beach side of the house was lined with sliding glass doors opening on to broad wraparound decks on both levels.

Armed with my smudgy drawing, I paid a visit to one of the local builders. The salesman who waited on me was either very inexperienced, very limited, or both. He seized upon the sketch as if it were a finished blueprint. Glossing over some of the obvious points, he said he would have the company architect give me an estimate within a few days on the basis of the drawing. True to his word, three or four days later he came up with a ballpark figure of $40,000. We realized the figure was meaningless since we hadn't discussed a thing about the interior layout or even how many rooms we wanted.

I'm not sure why we didn't go out and get an architect at this point. But all we wanted was a simple beach house, and "architect" sounded complicated and expensive. (As it turned out, the architect's fees proved to be among our more reasonable and manageable costs.) Instead we sought out another builder, a father-and-son team who seemed competent and reliable. We spent several weeks working on plans with the son, who agreed to just about everything we proposed. The problem was that he had no ideas of his own. When it came time to review the preliminary drawings with the father, we overheard him say, "But you can't move that beam, son. If you do, the roof will cave in." It was at this point that we decided we needed an architect.

C.B. thought it might be a good idea to write to the Yale University School of Architecture in the hope that they could recommend an architect. Our letter of inquiry met with an immediate response. On January 2, 1973, the phone rang and we were greeted by the deep tones of Don Watson introducing himself and saying he was responding to our letter to the university. That is how we first "met" Donald Watson, visiting professor at the Yale School of Architecture and a practicing architect. Little did we realize that he was soon to become our companion, confidant, planner, designer, counselor, confessor, and sometimes . . . formidable adversary.

Later in January, C.B. and I set up a meeting with Don at his office in Guilford, about halfway between New Haven and Pilots Point. His quarters are above a group of storefronts just off the village green. C.B. and I climbed the long, steep steps. As we got to the top landing, the first office door flew open. "You must be the Franks," Don began and gave us both a hearty handshake. His massive frame towered over C.B., who is six feet two. In spite of Don's somewhat serious demeanor, we found ourselves charmed by his warmth and friendliness. He put us at ease immediately by insisting on a first-name basis.

Don's office was crammed full of books and plans along with architectural samples and equipment. High wooden stools were his only seating. Soon he was proudly showing us photographs of his work—mostly schools, clubhouses and hospitals, but also a few private dwellings. They all had simple, bold contemporary lines; their interiors were open, spacious, and attractive. We liked Don's style at once.

Sensing our approval, Don drew us out with a few casual questions. Where was our building site? How large was it? Did we plan to live there year-round? How large was our family? How soon did we plan to retire? And, lastly, what kind of house were we considering?

By this time we had developed a few guidelines and preferences. The house was to be used year-round when we retired. We both liked contemporary architecture with its sharp angles, cathedral ceilings, exposed beams, and large expanses of glass. The living room would be very important to us and we'd like two or perhaps three bedrooms. We hoped the house could be on one level. In any event, the master bedroom should be on the ground floor. The house we had in mind was efficient and easy to maintain with simple designs and natural finishes. We hoped to get started as soon as possible.

When we finished this little briefing, Don turned to C.B. and asked if he had any special requests. C.B. said, "Not really. But I would like to be able to see the water from my bed. Oh, yes. And I'd like to have ceramic tile on the floors."

Don grasped our ideas quickly, nodding frequently as he scribbled some notes. He expressed interest in our educational backgrounds and wanted to know all about us. He drew C.B. out about his business and his interest in boating and computers, and he asked me about my painting and crafts and my career in psychiatric social work.

As we were about to leave, Don asked if we had any budget in mind. C.B. referred in an offhand way to the $40,000 figure the building salesmen had come up with and Don made no comment. We left our first meeting with Don greatly relieved, feeling satisfied that we were in competent hands at last!

On the drive home, C.B. suggested that I take over the planning of the house. I might have done better to heed the adage about fools rushing in where angels fear to tread. As a "professional" homemaker of many years standing and as an amateur artist, I did know something about homemaking and color coordination and I did have very definite ideas about what I wanted. But I might have been wiser to insist that C.B. take over while I sailed on a six months' cruise to the South Pacific. Then he could surprise me and I could fly home when the house was finished.

At our next meeting with Don, we adjourned to our beach site almost at once. It was dark and dreary that afternoon as we approached the Point, passing endless rows of deserted summer cottages sitting silent and neglected in the January chill. Ahead of us loomed the marina's large boathouses surrounded by bulky, canvas-covered hulls high on their wooden cradles.

We circled the marina's frozen basin making that last turn toward the beach. It is precisely at this point that we always experience a sudden exhilaration as the whole panorama of Long Island Sound suddenly stretches out before us. Our lot is only a few feet beyond.

As Don stopped at the site, he let out a long, low whistle. Without a word he bounded out of the car. Pivoting in wide circles, he paced back and forth as if he were marking the boundaries. Then he loped down to the water's edge. Just at that instant, the sun broke through the heavy clouds. The water sparkled and the land and sea stood out in sharp and colorful relief. Don was beaming as he slowly climbed back up the beach. We knew he shared our enthusiasm for this special spot.

By the time we met again, it was early February. This meeting was far less dramatic than the first two had been. Now that most of the preliminaries were over, we began to settle down to some serious business.

Don encouraged us to express even our wildest fantasies about the house. He already knew many of our ideas from our earlier meetings but he wanted me to elaborate on them. Don and I agreed that it would be very important to capture the wonderful seascapes wherever possible. Also, we wanted a living room large enough to entertain in but small enough to be comfortable and homey for just the two of us. In this house we really expected to *live* in the living room. (The living room of our Stamford house almost never seemed to be used by anyone except our dogs. We "lived" in the family room instead.) We'd had enough of formal dining rooms and formal dining but we did want an eating area separate from the "working" kitchen. And I was tired of running up and down stairs to monitor washers and dryers. This time I'd have the laundry room close to the kitchen and on the same level.

We were looking forward to the luxury of a large master bedroom suite with walk-in closets, and some division between our dressing room and the shower-toilet area. For Steve we hoped to provide a separate section of the house where he could entertain his friends in privacy. We wanted a two-car garage with plenty of extra space for C.B. to store his boating paraphernalia. We'd also need a basement large enough to house both C.B.'s workbench and tools and my craft equipment, as well as additional space for general storage.

Don seemed surprised that we wanted only two or three bedrooms. But we hoped to have Steve's den or sitting room double as a guest room. We didn't want the upkeep of empty bedrooms and, having previously owned summer beach homes, we were especially sensitive to the just-happened-to-be-passing-through syndrome. Anyone with a house at the seashore knows that while family and friends seldom impose, long-forgotten, casual acquaintances have an unerring instinct for tracking down empty beds at the shore.

It seemed that Don and I had just gotten started when he said we'd have to stop and continue our discussions the following week. By our next meeting, Don had produced some rough sketches and a tentative layout. One thing was obvious, he simply couldn't meet all our requirements on our tiny lot with a one-story house. There was only one way to go: up!

It was fascinating to see how many of our requests Don had managed to include even in these early drawings. I had expressed a list of preferences and Don had somehow turned my wishes into a house! The rooms were extra large and the house seemed to form a harmonious whole. We spent the rest of our appointment refining the plans, erasing a section here, adding something there. So far there were no windows, closets, or other details. Don certainly seemed to have a knack for carrying out our ideas, but there were times when he questioned them and would suggest more suitable architectural alternatives.

As our session drew to a close, Don handed me a printed folder, explaining that it was a standard AIA (American Institute of Architects) form for an architectural contract. He assured me it was quite routine but urged me to review it with C.B. and return it at our next meeting.

Until now, no money had changed hands and there'd been no mention made of any payment for Don. In looking over the papers, we weren't surprised to find that we would be billed on the basis of our hourly consultations with Don. The total figure was not to exceed 7 percent of the basic $40,000 budget for the house, or $2,800. (A very modest fee by any standards, even in 1973.)

Don's services would include designing the house and preparing the preliminary architectural plans. Later he would develop more detailed drawings and specifications, or "specs," as they were usually known. The builder would estimate his costs and bid for the job on the basis of the plans and specs. It was understood that Don would help us find a suitable builder and would act as our agent in overseeing all phases of the building contract. He would also represent our best interests in all matters relating to the house and see to it that it was brought to satisfactory completion.

Over the weekend C.B. and I went through even the fine print on the AIA form and decided to forward a copy of the contract to our lawyer before signing it. When we didn't hear from her for several days, we called and learned she would be out of the country for a week. In our eagerness to get started, we impulsively signed the contract before she had even a glimpse of it. Although I wouldn't recommend this rash behavior to anyone else, we were very lucky: Don proved to be totally honorable and reliable and we've never had reason to regret our hasty decision. Although later our building budget jumped to an astonishing amount, Don never raised his original fee.

Shortly after we signed the contract, Don specifically asked C.B. to come to our next meeting. It was at this meeting that Don first broached the subject of solar heating. Neither of us had ever heard of it. Had C.B. not been there, I doubt that we would ever have embarked on our solar venture. But as soon as Don mentioned solar heating, C.B. became very interested and immediately wanted to know all about it.

I was aghast. What the devil were they talking about? Here we'd owned the lot for almost three years; then we'd wasted months looking for a builder; and now C.B. and Don were talking about using some new heating method no one had ever heard of, and one which would probably delay us even longer.

I sat by fretting and stewing while C.B. tried to get Don to give him more details. Pleased with C.B.'s enthusiastic response, Don went on to tell us about Everett M. Barber, Jr., an engineering colleague at Yale who had been actively involved in solar research for many years. Barber, Don said, was convinced the sun's heat could be used economically and practically for domestic heating. Don lost no time saying he would arrange for us to meet with Barber in a week or two.

C.B. and I left Don's office that day in a state familiar to many married couples—C.B. was intrigued and stimulated by this new idea of solar heating, while I felt frustrated and somewhat incredulous of his attitude and behavior. C.B. couldn't wait to meet with this young engineer to hear more about his ideas for heating a house with solar energy. I could have willingly waited forever to meet this fellow and hear his absurd scheme. The whole concept seemed especially foolish here along the New England coast where a high percentage of the days are cloudy and overcast. Why couldn't we use oil or gas or electricity the way normal people did?

After a few terse comments, I lapsed into silence. Things will work out, I kept telling myself. Next week we'd meet Everett M. Barber, Jr., with his wild ideas about heating and C.B. would soon see the error of his ways. Then Don and I could finish designing the house using a regular oil heating system. Or so I thought.

If you're going to present a wild, outlandish scheme to someone who has already prejudged you and your ideas, it certainly helps to be charming. And Ev is that, not because he tries to be; he just is a charming fellow. C.B. and I liked him immediately.

When Don introduced us to Ev in March of 1973, we knew from his impressive credentials that he had to be in his thirties. But he seemed very boyish. In many ways, he is the opposite of Don. When they stand side by side, Ev's dark curly hair barely reaches Don's chin. Ev's brown eyes sparkle and his deep dimples add to his youthful appearance. While Ev is quick to laugh and has a strong sense of humor, Don tends to be far more serious and deliberate.

For all their dissimilarities, we soon learned that these two talented professionals provide an amazing balance for each other. Sharing a deep mutual respect, they are able to work together creatively and harmoniously.

After the usual introductions and exchange of pleasantries, Don sat back and let Ev take over. Ev gave us a brief rundown of his work with solar energy. He had spent many years both at the National Aeronautics and Space Administration and at the National Bureau of Standards. He was convinced he could develop a practical home heating system using solar energy—a system that could cut fuel costs in half!

Ev explained that the sun's energy was already being used in many different ways all over the world. As a matter of fact, he said, all our space satellites so far had been powered by solar cells. If solar energy had been used successfully in space programs, at least we knew that it worked. But solar cells are used to generate electricity, not to heat houses. Their success would not necessarily prove that solar energy could be used for heating homes.

Ev sensed our growing interest and launched into a lengthy and dynamic presentation of his favorite subject. He assured us there was nothing mysterious or visionary about the use of solar energy for heat. Solar water heaters had been in use in many parts of the world for more than fifty years. They were common in Australia, Israel, and Japan, to name only a few. Solar water heaters had once been popular in this country too, during the thirties and forties. But as power became cheaper, people lost interest in them.

Ev explained there were many other ways in which the sun's energy had been utilized. Solar furnaces could be used in industry and could reach incredible temperatures. One large solar installation in the French Pyrenees had reached a temperature of 3,500°F. This was done, Ev went on to say, by focusing large amounts of solar radiation on a special area. In some parts of the world, especially in developing countries, solar purification and desalinization of water were quite common. And solar energy was also used in crop drying and for powering irrigation pumps.

Ev seemed to be opening up a whole new world for us. This wonderful source of energy had always been available to everyone. And it didn't need to be drilled, mined, piped, shipped, or generated. It was clean, nonpolluting, came in unlimited quantities—and was absolutely free! It was hard to believe it had been around since the beginning of time and that we had always taken it so completely for granted.

It was obvious Ev Barber had given a lot of time and thought to developing a solar heating system for domestic use. After leaving his government position, he had continued his solar studies and was closely involved with national and international organizations actively promoting the use of solar energy. He assured us that dozens of American scientists were currently studying and experimenting with solar energy. In recent years many of our large universities in Arizona, California, Colorado, Florida, and Wisconsin, to name just a few, had launched impressive solar research programs. Other solar demonstrations were under way in different parts of the country.

Ev was convinced our energy shortage would become much more serious in the next few years. He felt certain the days of cheap power would soon be gone forever. Our fossil fuels were being used up at an alarming rate and, he believed, would sooner or later be entirely depleted. Unless we soon developed some new kinds of energy, Ev pointed out, the United States would face a critical shortage of energy and would find itself unable to maintain its high standard of living.

Although Ev's presentation was clear and logical, we still could not understand why solar energy had not already been used more extensively for domestic heating. Ev told us that some efforts in that direction had been made in the thirties and forties when several experimental solar houses had been built in Massachusetts. Although he was sure these experiments had been successful, he couldn't tell us exactly what had become of those solar dwellings. But he thought it would be easy enough for us to check on them. All but one of the houses, he said, had been built and funded by the Massachusetts Institute of Technology (MIT).

Despite many unanswered questions, C.B. was obviously very stimulated by the whole idea of using the sun to heat our new home. He wanted to know just how soon Ev could draw up a heating plan for us.

Delighted with C.B.'s response, Ev said he would get busy designing a system right away. He had no idea how long it would take to work out the details, but he would meet with Don and get back to us with some facts and figures as soon as possible.

C.B. offered to make an advance payment on the heating system but Ev assured him that wouldn't be necessary. At the beginning, his work would be largely theoretical and would require no immediate outlay of funds.

By now everyone was feeling more relaxed. There was much smiling and shaking of hands. This was the magic moment when our solar house was actually conceived.

Our trip back to Stamford that day was much more pleasant than our last trip had been, but I have to admit that I still had my reservations about the idea of solar heating. Again, quite the opposite was true of C.B. He liked Ev Barber and the way he went about explaining things. And he was impressed with Ev's background and experience. C.B. wasn't at all concerned that there was nothing in writing about the heating arrangement. He was used to doing business on the good faith and mutual trust of the parties involved. He felt strongly that no written contract was half as important as the ability and integrity of the contracting parties. He never doubted (then or at any future time) that both Don and Ev were honest, competent, and dedicated practitioners.

My enthusiasm for this new and untried heating plan was still less than wholehearted. I was surprised that C.B. was willing to jump into this so quickly, knowing so little about it. After all, Ev was a complete stranger. It was just possible this whole heating idea could be some impractical, fly-by-night scheme.

C.B. and I decided to find out what we could about those early Massachusetts solar houses. And what better way than to speak to a man who owned one? But, try as we might, we couldn't learn very much (at least not at this time). C.B. called several different departments at MIT but no one seemed to know anything definite.

I decided to try our public library. However, even our normally well-stocked local library had very limited material on solar energy and no references at all to the MIT houses. Scarcely a handful of titles appeared under the solar energy heading and most of them had been written many years earlier. Surprisingly, one architectural volume talked about what has since become known as passive solar heating. It explained the importance of having as many south-facing windows as possible so that sunlight can stream directly into the building.

Other books were so theoretical or technical I could not begin to understand them. Several volumes treated the subject simplistically—or so I thought at the time. They gave directions for backyard and classroom experiments, reminding me of the old Boy Scout routine of lighting a fire by concentrating the sun with a small piece of tin or glass. It didn't occur to me then that this, too, is a good example of solar power. The less technical science magazines had a few references to solar energy, but these were usually brief and superficial, reporting largely inconclusive results.

I continued to work with Don on the house plans. Meanwhile Ev was working out the mechanical details for our solar heating system. Somehow, although Don and I never discussed the decision to use solar heat, or how this might affect the design of the house, we were now viewing the plans and the house in an entirely new way. No longer were we building an ordinary dwelling. Rather, we were planning a very remarkable house—a house to be heated by the power of the sun.

Several factors made our proposed beach house a prime candidate for a solar experiment. For one thing, our south-facing location would give us the necessary continuous exposure to the sun. And with a 100-foot-wide beach stretching southward from our lot, it was most unlikely that anyone or anything would ever obstruct our freely flowing sunshine. The open waters of Long Island Sound, extending ten miles in a southerly direction, would give us complete protection from any possible interference with all that powerful sunshine beaming down on us.

Don and I had always agreed on one primary objective: to capitalize on the many lovely sea views whenever possible. Since the Sound was to the south of us, it was only logical to put all the large glass-walled living areas on that side of the house. This coincided with one cardinal principle of good solar design: maximum glass exposure to sunlight on the south side of the building. Don's design was also in keeping with a second fundamental solar rule; that is, to have as few openings as possible on the north side. This side of the building is usually the coldest since it gets no direct sunlight. Also, New England storms and winter winds frequently come from a northerly direction.

Adding height to our house proved to be another solar blessing. Upward expansion meant a smaller roof and therefore less chance of losing heat through that surface. A compact house with several different levels would obviously be more energy efficient than a one-story elongated ranch. A higher house would also decrease the possibility of ever having a taller nearby building block off or interfere with our direct exposure to sunlight.

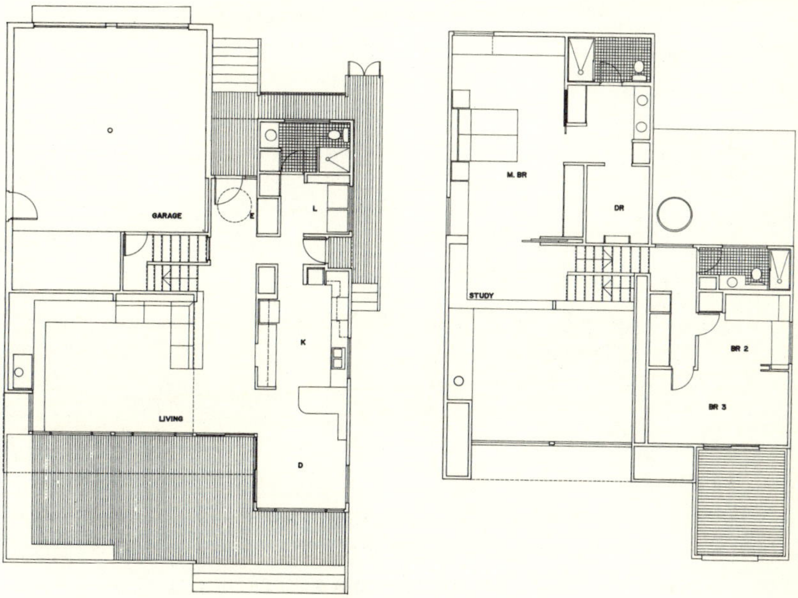

1 2

Courtesy of Donald Watson, AIA; from *Designing and Building a Solar House*, Gardenway, 1977

Sol-A-Sea: First-floor living area; second-floor study and sleeping areas

Don's plans were almost a perfect example of passive solar heat collection; that is, taking advantage of the sun's direct rays without the aid of mechanical devices. The sun would pour into our 22-by-16-foot living room through the oversized windows and sliding glass doors running across the entire south side of the house. During the day the heat would be trapped inside the house with its three glass walls; that is, far more heat would be retained *inside* the walls than would be lost *through* them. And, also in keeping with sound solar practice, Don had recommended the use of double-glazed windows to minimize heat loss.

Our dining area, jutting out some 10 feet beyond the rest of the house and with three glass walls, would be a natural solarium. Morning sunshine would flood in from the east, followed by continuous sunlight infiltrating from the large south windows. In the afternoon the sun would come through the sliding glass doors on the west side. This arrangement would also create a perfect buffer between us and our neighbors to the east and would protect our privacy both in the living room and on the deck.

The dining area would be divided from the kitchen proper by a 3-foot wide, table-height counter which would double as a breakfast table. With such a low divider, there would be nothing to obstruct the free flow of light and heat from the dining area into the kitchen. Also, we'd have clear views of the water while doing our kitchen chores. There would be one east-facing window centered over the sink. From here you would be able to see many miles across the water in the direction of Long Island's north shore.

With our open floor plan, the main living areas on the first level of the house would have full advantage of our exposure to the sun on the south and east. A long (28-by-5-foot) entry hall, running through the house from the north-facing street entrance to the sliding glass doors on the south side, would also serve as a divider between the living room and the dining-kitchen area. There would be a small pass-through between the kitchen and entry hall to make kitchen-living room communication easier.

Both the front entry and the back entry on the east side of the house would be recessed several feet to provide extra protection from the elements.

A small (5-by-6-foot) alcove between the kitchen and the downstairs guest bathroom would serve as our laundry room. A bifold door could close off this section when it wasn't in use. The large stainless steel sink in the guest bathroom could also serve as a washtub. Two small, high, east-facing windows would bring in morning sunshine and provide fresh air for drying clothes.

The large (10-by-6-foot) guest bathroom beyond the laundry would also have two small, high windows; these would face north on to the street. They would bring in continuous daylight without admitting any direct sun. Most likely, the bathroom would be colder than other parts of the main living area on this level. But since bathrooms aren't normally in continuous use, it is less important to keep them warm all the time.

The bedrooms would be contained in the top two living levels. Up seven low-rise steps, about a half-story above the first level, would be a large (14-by-6-foot) balcony overlooking the living room and with a magnificent view of the water through the top of our high living room windows. C.B.'s desk would be built in this area, which would double as his study. An open railing would be the only barrier between the balcony and the living room, thus allowing both light and heat to flow up to the balcony and into the sleeping quarters.

Our plans also called for the use of several pocket doors on the upper levels. When not in use, these doors slide into the side walls, allowing light and heat to circulate freely. When in use, they provide privacy and separation and conserve heat as well.

Along the balcony's 12-foot ceiling would be two clerestory windows bringing in a steady north light. We'd often admired windows like these. They tower high above other parts of the interior, sometimes rising above the roof itself. They have a contemporary look and bring penetrating shafts of light deep into interior spaces. They convey a feeling of closeness to the sky and to nature itself and their height allows for complete privacy without the necessity of shades or draperies. Clerestory windows are energy-savers too. The added natural light means less need for artificial light and can result in real savings in electricity. Several Clerestory windows were included in our plans from the beginning, and, as things turned out, they form a fitting and attractive balance for our rooftop solar collectors.

The master bedroom suite would rise over the two-car, ground-level garage. The large (23-by-13-foot) sleeping chamber would look out to the Sound through the top of the living room windows, fulfilling one of C.B.'s two basic requirements for the house, i.e., that he be able to see the water from his bed. Later we would build C.B.'s chest of drawers in the master bedroom under a west-facing window, one of only two such windows in the entire house. Normally in a solar house, west-facing windows are kept to a minimum because of the hot afternoon sun in the summer months. However, we did want to take advantage of the spectacular views to the west. And since the prevailing winds are from the southwest, they provide a delightful cooling effect.

A large double awning-type window directly beneath one of the clerestory windows in our bedroom would provide one of the few northerly views from the house, looking out over the street and beyond toward the marina. (Later we would make plans to build my desk in this location, providing me with a small study area.) The four clerestory windows on the north wall would effectively combine to provide excellent light and cross-ventilation throughout this second level.

Separated from the master bedroom by another pocket door would be a 7-by-8-foot dressing room with two small, high, east-facing windows. Next to that would be a 9-by-10-foot walk-in clothes closet, the only area in the whole house without direct exposure to natural light and air. Because this small space was so sheltered, it later proved to be a perfect refuge from storms for me and Brandy, our German shepherd. For all his ferociousness with strangers, he is terrified of storms. When the elements become too frightening, Brandy and I hide out in the closet together.

Our dressing area would also have a double sink with vanity. Next to it would be our bathroom with one north-facing clerestory window.

To insure everybody's privacy, our son Steve's rooms would be located on a third level, seven steps above the balcony and above our master bedroom suite. This whole area could be shut off from the rest of the house by another pocket door at the top of the stairs. Opening off a 14-foot hall on this level would be a full bathroom with two small, high windows facing north. The hall would lead to Steve's compact (10-by-8-foot) sleeping quarters, just large enough for his captain's bed, a desk and a chest of drawers. This area would be separated from a 10-by-13-foot sitting room by a folding door. Either of these rooms could be reached directly from the hall. By closing the folding door, we could always use this sitting room as a guest room. And when Steve was away at school, we could use this room as an upstairs den or study. South-facing sliding glass doors off the sitting room would open on to Steve's private 10-by-12-foot deck. The views from here would be the most dramatic of all—with the full sweep of the Sound looking south and east toward Plum Island and Gardiner's Bay. It would be a perfect, private retreat for sunbathing, reading or just relaxing.

There it was, at last, and finally, our solar house. But it was all on paper. It would still be many months before our house would be life-size, before we could step into it, walk from room to room, and enjoy its beautiful views.

Although I had been impatient and often found our progress much too slow, we had, in little more than three months, actually planned and designed our solar house. To translate our plans into reality would, however, involve many new and unfamiliar kinds of equipment and methods for which we weren't entirely prepared. Another complication was the growing energy crisis. While we wrestled with house plans and construction, energy shortages were becoming more of a problem for all Americans. Little did we realize then that eventually this scarcity would disrupt our economy and impinge upon all of our lives.

So far, efforts by Americans to conserve gas and oil and to free ourselves from dependence on imported energy were inadequate. We were continuing to import more oil from the Middle East each day, each week, and each successive month. While the price of oil continued to rise, the economy began slowing down. Supplies of gas and oil were dwindling while the lines at service stations were growing. Those of us who had lived through the blackout of 1965 knew only too well how a massive power failure could create havoc with our economy and our everyday life. There seemed to be no easy answer to this growing crisis. And in 1973 the government was exerting little leadership, continuing to rely solely on voluntary cooperation to meet these energy problems.

With all this going on around us, solar heating was looking better than ever. My earlier fears and doubts about "going solar" were slowly disappearing and I was starting to realize how lucky we were to be able to take advantage of this "new" kind of energy just when other forms of energy were becoming scarce. Solar heating no longer seemed like an interesting experiment. It was looking more and more like a practical way for us to meet our future energy needs.

2

Plans and Specifications

When we next met with Ev Barber, we brought with us an extensive list of questions about our solar heating system. Exactly what kind of equipment would we have? How would the heat be regulated? How much maintenance would be involved? How long could we expect the equipment to last? Who would service the system and replace its parts? What would happen if the system failed in the middle of winter? How would we keep warm when there wasn't any sunshine? And, of course, how much would all this solar apparatus cost us?

Fortunately for us, Ev Barber has a knack for explaining complicated ideas without using a lot of confusing or technical terms. The way he explained it, solar heating sounded quite simple.

To make use of the sun's heat, we would first need to collect it. The collectors would be the heart of the entire solar heating system, capturing the sun's energy and putting it to work. The collectors would be small metal boxes or troughs which would trap the sun's heat as it passed through their glass covers. Within the collector, the solar energy would heat a special fluid flowing through a series of coils embedded in the collector's base or plate. The coils would pass from the collectors to and through a large storage tank in the basement, heating up the water in the tank. The heated water would either be stored in the tank until needed or run through coils in the blower unit, thus heating the air to be blown into the living spaces—much as any forced-hot-air heating system works. At this time Ev wasn't sure exactly what size this tank would be but he knew it would be "very large."

C.B. wanted to know if there was any chance that the liquid in the collectors would freeze during our cold New England winters. Ev explained that he would be using a special propylene glycol solution for this very reason; it would be somewhat similar to the permanent type of antifreeze used in car radiators.

C.B. and I had found it especially difficult to understand what would happen on those cloudy, rainy, and snowy days so prevalent during Connecticut winters. When the sun wasn't shining, just how would we get enough heat to keep us warm? While we knew there was plenty of strong sunshine in places like Arizona and California, how much, how strong, and how reliable was New England's winter sunshine? Besides that, how is sunshine measured? Obviously you can't weigh it or figure it by the yard or gallon.

Ev assured us there were accurate ways of measuring sunlight, or solar radiation as it is called. Actually, he told us, there are two kinds of radiation. One kind comes to us directly from the sun and is known as direct radiation. It is usually visible as bright sunshine. The other type, known as diffused radiation, is scattered through millions of particles in the atmosphere and is often barely visible as sunlight coming to us through clouds, fog, or haze. Diffused radiation is very important too, because it adds greatly to the amount of overall radiation which can be collected and used for heating purposes.

Ev used still another term for the heat and light we receive from the sun: solar insolation. This is usually measured in units called Langleys, a Langley being equal to one calorie of radiation per square centimeter. Langleys are used to calculate the amount and intensity of solar radiation falling on any surface at a given time.

With over three hundred weather stations operating across the country, it would not be difficult to find out the exact amount of solar radiation we could expect in our area during the next heating season. Further, the National Climatic Center in Asheville, North Carolina, has been keeping detailed records of weather conditions all over the country for years. The center has data on temperatures, precipitation, and winds, as well as exact calculations on the amount of sunshine we could expect to find in different areas. While there would always be minor variations in individual locations due to local altitude, wind, fog, and other atmospheric conditions, this basic weather information would be essential in designing our solar units. There were day-by-day, month-by-month, and year-by-year statistics readily available.

Additional and more technical data, Ev said, is available to engineers from the American Society of Heating, Refrigerating and Air-Conditioning Engineers. ASHRAE can supply detailed reports, charts, and graphs on weather conditions through its New York headquarters. Thus Ev assured us he did not anticipate any difficulty in collecting all the technical information he would need for planning our solar heating.

To design our heating system adequately, Ev would have to study the records and determine our average winter temperatures over the past ten years. In this way, he would know generally what kind of temperatures we could anticipate in future winters.

As Ev explained it, daily temperatures during the heating season are averaged and expressed in "degree-days." While we had often heard degree-days mentioned on television weather forecasts, we were never sure exactly what they were.

Ev, with his special gift for teaching, clarified this using easily understood terms. Degree-days, he said, are based on the assumption that indoor heat will be used whenever the outside temperature falls below 65° F., Take the average outdoor temperature for one day and subtract it from 65. The difference or remainder is the degree-day for that particular day. For example, if the average outdoor temperature on a particular day was 40° F., you would subtract 40 from 65 and come up with a 25-degree-day figure. Once we realized that a degree-day wasn't a "day" at all, but an expression of the difference between a norm and the actual temperature, the concept was easier to grasp.

The total number of degree-days for one heating season would give you a fair idea of the number of degree-days to expect in future heating seasons. Naturally, degree-days vary across the country. While North Carolina might have 4,000 degree-days during a given heating season, Minnesota might have twice that number.

According to Ev's calculations, the heating season on the Connecticut shoreline would probably average 5,000 degree-days and our heating system would be designed on that assumption. These calculations, Ev said, were critical for designing a heating system of maximum efficiency. An oversized system would be wasteful and an undersized system would be inadequate.

There were many other things about solar heating that we still did not understand. How large would the collectors be? How would they be attached to the roof? Where could we buy collectors?

Ev hoped to do most of the work on the collectors himself. There was literally no place where collectors could be bought in the United States since they were not yet being manufactured here. He hoped to get some assistance from his colleagues and from some vocational teachers in a nearby high school.

Ev said he would probably use copper sheeting and tubing for the backplate or absorber of the collector. The copper coils for carrying the heated glycol liquid would be soldered to the backplates, which would be coated with a special kind of absorbent black coating. As Ev explained, dark colors absorb heat rather than reflect it; but light colors tend to reflect heat and light rather than absorb them. Ev had been developing a special kind of black coating which he referred to as a "selective coating." He hoped it would markedly increase the collectors' efficiency in gathering and retaining solar heat.

The hand-fabricated collectors would have to be expertly fashioned so that there would be as little solar heat lost as possible.

We were still having difficulty visualizing the collectors. Sometimes Ev referred to them as boxes or panels, while other times he called them troughs. We knew they were to be attached to the roof somehow, but we couldn't imagine how or where this would be done, or if they might change the appearance of the house.

By now we did understand that the collectors would have to face south in order to collect as much solar energy as possible. According to Ev, they would also have to be slanted at a specific angle.

We were now aware that in addition to our solar heating system, we would need some kind of auxiliary or backup supply of heat. Ev didn't think it would be feasible (or economically sound) to design a solar system that would supply all our heating needs. To do so would have required an extraordinary number of collectors covering an enormous area (more than four times as large as the system he had planned for us). Our roof simply would not be large enough (or strong enough) to hold that many collectors. Even if it were, the cost of such a large number of collectors would have been prohibitive. Consequently, our solar system was designed to meet only slightly more than half our estimated heating needs. However, with the skyrocketing price of oil, even this could represent a savings of hundreds of dollars in fuel costs each year.

The auxiliary system would be needed to back up or supplement our solar heat during prolonged periods of cloudy or sunless days. It would also be necessary to supplement the solar heat during extended spells of very cold weather. The solar-heated water in our storage tank would probably be enough to keep the house quite comfortable at night and during one or two rainy or cloudy days during mild winter weather; beyond that, if the sun was still hiding or not strong enough for us to collect more solar energy, we would probably have to turn to our supplementary source of heat. Since our auxiliary heat would not be used very often or continuously over long periods of time, a small conventional hot water heater could serve this purpose. Ev assured us that such a unit would be able to supply an adequate amount of heat when necessary.

Once the solar heat was collected, Ev said, it would be distributed and put to use in much the same way as any forced-hot-air heating arrangement. There would be a large fan or blower to send the solar-warmed air through the ducts to be blown and circulated throughout the house. Because this equipment would not require any special parts, the installer would not need any special skills or previous experience with solar heating. Any good local plumber would be able to install and maintain our system once Ev briefed him about its operation. Ev estimated that our solar components should last at least twenty-five years, although he couldn't guarantee this in writing.

Originally Ev had told us that he couldn't give us an accurate cost estimate of the solar system for some time, but C.B. was anxious to get some idea of what all this would cost. Ev came up with the figure of $3,500 for all the solar parts. He estimated that this was roughly twice the cost of a conventional heating system at that time. In other words, the entire heating apparatus, including both the solar and the auxiliary systems, would cost about $7,000.

C.B., who had taken over responsibility for this aspect of the planning, seemed to feel this figure was reasonable. But he wondered if Ev knew approximately how much money solar heat could actually save us in fuel costs. Ev said that although a good deal would depend on the price of fuel when our system became operational, he was thoroughly convinced of one thing: the price of oil and other fuels would continue to rise as they became scarcer. The cost of electric heat was already very high and Ev was certain this would continue to go up. Although natural gas still seemed plentiful and was relatively cheap, he believed that future gas shortages were inevitable. Ev pointed out an important consolation for us: The higher the cost of conventional fuels, the more we stood to save on solar heating.

If our solar heating was able to supply between 45 and 55 percent of our heating needs, Ev felt sure we would be able to save enough in fuel to recover our capital investment in the solar equipment within the next ten or, at most, fifteen years. (As things are turning out, we will probably recover our investment in less than ten years.) But since this was uncharted territory, Ev could give us no assurance at this time of the actual payback period.

For the first time I began to understand what C.B. meant when he said how important it was to trust people and have faith in those with whom you have business dealings. C.B. had become very fond of Ev and had a great deal of respect for his ability. He trusted Ev implicitly and a kind of father-son relationship of mutual affection and admiration was developing between them.

While Ev was still working out the plans, we had no written contract for the heating system. As yet, there was nothing on paper; nor had one cent passed between us. C.B. felt the worst that could happen was that we might lose the $3,500 capital we would be investing specifically in the solar equipment. If the solar heating didn't work the way we hoped it would, we would still be able to use the auxiliary system as our main source of heat. C.B.'s innate optimism and basic trust in other people—along with his natural attraction to innovation and adventure—had already converted him into a solar enthusiast. Without these qualities and C.B.'s consistent support and belief in Ev and his ideas, we would certainly never have completed our solar venture.

During this period, I was totally preoccupied with finding and selecting materials and equipment now that the design of the house was almost complete. Until this point I hadn't realized how many decisions must be made before the architectural plans can be submitted to a builder. Builders need to have the exact specifications—"specs"—before they can estimate the costs of construction. They would need a full accounting and description of all the equipment and materials necessary for carrying out the details of the architect's drawings and stipulations. For example, the specs would need to describe the exact quality of the steel, wood, and masonry to be used, as well as the full particulars about the external siding and interior woodframing, and the criteria for the standards of carpentry and craftsmanship that would be expected. Even minor details, such as the kind of nails to be used, would have to be included. (Actually, the latter is an important consideration in our damp seashore location where everything rusts so easily.)

In addition, the specs would have to outline the number, measurement, style, and quality of all doors and windows; the grades of Formica, fiberglass, stainless steel, wood, and other materials to be used; and the kind of paint, stain, or other finishes to be applied on all exterior and interior surfaces. Finally, the specs would state how the builder should grade the site and finish the driveway and walkways once the house was completed.

Basically, the specs for our solar house would not differ substantially from the specs for a conventional home. There were, however, some important differences. As with most conventional houses, our specs called for 3½-inch batt-type fiberglass insulation on all the side walls. But, unlike most conventional houses, ours would also have an inch of rigid cellular polystyrene insulation on the outside of the plywood sheathing, underneath the wood siding. Other items would probably be new and unfamiliar to most builders, such as a silicone-coated fabric roof. (This idea was later abandoned.)

The solar heating system itself, the so-called mechanical plan, was not to be included in the specs or submitted to any builders at this time. Unfortunately, it was this very unusual situation which led to complications later on. In retrospect, it is clear that the circumstances were unavoidable because everyone was working without the guidelines of precedents. Ev was still doing research and developing the solar heating plan long after the design of the rest of the house had been completed.

Also, while much of the plumbing would eventually be installed by the builder's plumbing crew, the plans for the solar heating unit would have to be worked out first. Once these plans were finalized and approved, Ev would undoubtedly be working closely with the builder and his plumbers to finish the standard plumbing system and to help with the installation of the solar equipment.

As our architect, Don would be deciding on many items in the specs without needing to consult us. These would include the heavy construction materials like concrete, steel, and the basic lumber, as well as some of the specific processes to be used, such as damp-proofing, caulking, and flashing.

There would, of course, be a wide range of choices for us to make with Don's help, such as the selection, quality, design, and color of many other materials and products. One of the first decisions we would need to make concerned the exterior siding. Don and I, with C.B. concurring, had agreed earlier that wood siding would be ideal; it would give the house a natural appearance and would be especially well suited to our beach setting. Since wood is an excellent insulator, the siding would be another solar asset.

At our next conference with Don, he mentioned having access to Atlantic white cedar, a special kind of siding unavailable from local suppliers at the time. It grew in swamps and for this reason alone would be ideal for our damp seaside climate. It could be used in its natural state without any paint, stain, or preservative and in time its raw appearance would weather to a silvery gray. Don suggested that we first take a look at a local architect's house covered with this kind of cedar before making a decision.

The house was about three years old and the siding had already bleached out to a pale driftwood tone. I thought it looked bleak and stark, but I had to agree that it would be especially appropriate for the design and location of our home.

So it was that on May 25, 1973, we became the proud owners of "4,000 board feet of kiln-dried Atlantic white cedar with tongue and groove"—our very first purchase for our beach house and an expenditure of $2,500 which assured us that the lumber would be available when needed. Don arranged to store the siding without charge in a friend's barn in Guilford. At this time we didn't realize that Don had saved us a good deal of money. Several months later, when we reordered additional siding to panel the living room, we found that it was even more difficult to get and the price had gone up enormously.

The cost of lumber, like the cost of many other building materials in the spring of 1973, was at an all-time high. Although it didn't seem to be the best time to build, inflation was continuing its relentless march and it seemed unlikely that prices would be going down anytime in the near future. C.B. and I agreed that we should finish the plans and find a builder as soon as possible.

Although many materials to be included in the specs wouldn't be needed until the house was under construction, they would have to be chosen now so the builder could estimate the costs. Also, many supplies not included in the specs would have to be ordered well in advance of construction.

By this time, deliveries were often held up many weeks due to a general slowdown in production and transportation. The large quantities of fiberglass and Formica that we'd soon need were not likely to be on hand or readily obtainable. The shortage of oil and gas was already affecting the shipment of these and many other building supplies. Some of the substances we needed turned out to be petroleum by-products and were either scarce or were rapidly becoming totally unavailable.

High on our list of priorities would be choosing windows and doors. Although important and very basic in planning any building, these openings proved to be of special significance for our solar house. We were now aware that while windows are an excellent way of trapping the sun's heat, they can also account for as much as 50 percent of heat loss. The quality and fit of the windows and doors in our solar house would have to be given top priority.

Even though the size and location of our windows and doors were already included in Don's drawings, we would need to choose the exact style and framing. These decision swould affect not only the appearance of the house and the views, but also our comfort, convenience, and consumption of energy.

Over the years we had always admired sliding glass doors—maybe because we'd never had the luxury of enjoying them ourselves. They seemed especially suitable for a beach house where outdoor living is so important. Actually, my original rough sketch of the house had included a series of sliding glass doors stretching across the entire beach side of the house.

I had assumed that both the sliding doors and the windows would be framed in wood. I had looked into this quite thoroughly, talking with several friends who'd lived on the water for years. They were unanimous in advising against the use of metal framing. Metal is a poor insulator, conducting both heat and cold. In the summer the frames would be hot to touch; in the winter they would be icy cold. But more important, they would also function poorly in keeping the cold *outside* the house and the heat *inside* the house. Several of our friends also pointed out that at our seaside location the metal frames would be particularly prone to corrosion and to pitting by the blowing sand. These factors might combine to damage the metal frames in a relatively short time.

When I presented my ideas for using wood window frames and for the installation of several sliding doors, Don flatly rejected them. In a way, this conflict was the opening salvo in a long battle of wills that many architects and clients probably engage in—the architect with sound practical ideas reflecting years of training and experience, and the client with an overpowering urge to realize his or her dreams and fantasies of the perfect house.

C.B. and I, like many couples, had never built a house before. We'd spent our entire married life in other people's dream houses which never quite suited our taste or met our needs. Now we wanted our very own house—something unique and tailored to our way of life. The trouble was that Don, quite understandably, must have wanted the house to be his, also. After all, it would be the product of his creative efforts and design.

The tug-of-war that started over the windows and doors had a set of conflicting themes behind it that reappeared throughout our subsequent encounters. There I was, armed with my hopes and dreams, a strong aesthetic sense, and the knowledge that

C.B. and I would not only be paying the bills but would also be living in the house. And there was Don, fortified with his years of architectural studies and his successful practice, his strong aesthetic sense, and the knowledge that his name would always be associated with a house that could—and did—become a prototype for solar-home building in the Northeast.

So there we were, squared off for battle. I marshaled all my arguments on the windows and doors and the appropriate framing. But none of them seemed to hold up in the face of Don's opposition.

It was only much later, when the heat of the controversy had dissipated, that I came to understand Don's sound reasoning for limiting the number of sliding doors. By substituting windows for most of these doors, we would have a tighter, more secure house. The windows would undoubtedly fit more snugly than the doors and would probably provide much better insulation and protection from the elements. Also, by having only one set of sliding doors at the beach end of the entry hall and one set in the dining area, we would be setting up a desirable traffic pattern: people could go to and from the beach, using the center hall as a passageway rather than cutting through or across the living room—a significant factor, considering the number of wet and sandy feet that would be trekking in and out of these portals for many a summer to come. We would also gain several feet of space beneath the windows, making it possible to have front-row views of the scenery without fear of drafts. An even more obvious advantage of windows over doors is the security factor. The latter would make the house less vulnerable to intruders.

I reluctantly agreed to have only two sets of sliding doors, each being 6 feet wide. But I wasn't about to acquiesce on the matter of wood framing for the windows and doors. I stuck to my guns and Don and I soon found ourselves at loggerheads. Later I decided to resort to a strategy that I'd found successful in some other tight situations. I composed a note to Don with great tact and care. Starting out with some innocuous pleasantries, I went on to my enthusiastic approval of the progress we'd made in planning the house, slowly and subtly leading up to the area of conflict. At first meekly and gently, I stated my objections, gradually building up to a final, mighty crescendo: The whole idea was intolerable, unacceptable and would have to be changed!

In this initial skirmish at least, the tactic brought some success. Don agreed to do some additional investigating about the window frames and soon came up with a solution. He found a new product just on the market, a wood window clad in painted metal that had a lasting, baked-on finish and a thermal break. This would give us the best of both worlds. At our next meeting Don brought in some samples, and happily, we both agreed that this wood/metal framing would be a perfect compromise. Eventually both sets of sliding doors as well as our windows were framed in this way. (This attractive and practical framing added very little extra cost. It has maintained its appearance and held up very well through the years.)

Now that the design and location of the windows and doors was settled, we would need to plan for many interior fittings and equipment. We chose fiberglass fixtures for all three bathrooms. Since the shower and tub units came molded into one piece and were too large to get through an ordinary door, they would have to be ordered and installed before the house could be completely framed in. Fiberglass is tough and durable and requires little maintenance, making it ideal for plumbing fixtures. Since the bath units were six feet high and would completely surround the area, there would be no need for tile. We'd never have to worry about regrouting or replacing tiles. Also, the fiberglass would be easy to keep clean. It had an unusually smooth and attractive surface and came in almost the same variety of colors as most standard porcelain plumbing. For much the same reasons, we chose Formica wall paneling rather than tile for the bathroom walls, an innovative and practical use for this durable material, which also requires little upkeep.

Up until now, we'd been so preoccupied with other decisions that we'd given little thought to any color scheme. But since we were beginning to order many materials and fittings, we'd need some basic color plan for the entire house.

As we mulled over the infinite variety and untold combinations of colors, we found ourselves unconsciously developing a kind of solar aesthetic. Even at this early stage, everything relating to the house made us more conscious of our environment—the clouds, the wind, the tides, and that all-important fiery sphere, the sun.

With a backdrop of sand and sea, it seemed elementary that we bring our beautiful surroundings into the house. What better way to accomplish this than to use the lovely blues and greens of the sky and the sea, blended with the warm beiges, golds, and browns of the sand and earth?

We planned to use one basic theme to bring the house together and to smooth the visual transitions from one space to another. With the interior plan so open, and with few solid walls between rooms, the colors—like the light and heat—could flow freely from one section to another. Sitting in the living room with the dining room in full view, you could look through the small pass-through into the kitchen. When the pocket door was open on the balcony, large sections of our master bedroom would be visible from the living room. And the full sweep of the open stairwell could be clearly seen from all three levels.

To avoid the monotony of the same colors in every room, we planned to vary the intensity of the shades and tones from room to room. In this way, we could combine the same basic colors in a variety of tones, accents, and highlights, making each room unique yet adding up to a pleasing and harmonious blending of color.

At this time, before plans and specifications went out to bid, we would need to decide on floor coverings for the whole house. Although it would still be many months before any flooring could actually be installed, our choices could make a significant difference in the builder's bid. For example, if we planned to carpet the house wall to wall, the builder would be expected to provide only standard subflooring at minimal expense and we would buy the carpeting. But if we planned to use vinyl or composition flooring in the kitchen, halls, and other parts of the house, the builder would be expected to supply these and would include this cost in the estimate.

Another alternative, and one high on our list, was the generous use of ceramic floor tile to carry out C.B.'s original request. In this instance the builder would need to include the additional expense for ceramic tile. Because this kind of material is heavy and could add appreciable weight, we found out later, the builder would probably need to prepare the subflooring in a special way. This could again add to the overall cost. Another expense would be the installation of the tile. Each piece would have to be grouted and set by hand. Even back in 1973 the services of a highly skilled tile-setter could be purchased only at a premium price.

It was a refreshing change for both Don and me to find ourselves in thorough agreement on at least one issue. There wouldn't be any problem about the flooring since we were both equally anxious to comply with C.B.'s wishes. And we soon found that ceramic tile had many fine qualities to recommend it for a beach house—most especially for a *solar* beach house: Tile is easily cleaned with soap and water and never has to be waxed; sand can be swept out without scratching the surface of the tile and, best of all, the tile itself would actually contribute to our solar efficiency. After absorbing the solar heat pouring through our south-facing glass walls all day, the tile could store or retain some of the heat for our nighttime use. We were also agreeably surprised to find that the cost of moderately priced ceramic tile compared favorably to top-of-the-line no-wax vinyl flooring.

I had a difficult time finding the right tile for our house. In some instances, tiles weren't being manufactured; in others, shipments were sporadic and subject to long delays. I finally discovered a large, well-stocked distributor near Danbury, about a half-hour's drive from our home in Stamford. When I found some attractive and acceptable tile samples, the proprietor generously offered to let me take them home. Happily, C.B. approved of my choices. Unhappily, Sunday morning's newspaper carried an ugly headline: "Large Tile Warehouse in Danbury Leveled by Fire"! Why did the fates go to such extremes to discourage me? All I wanted was to build our seaside house and live in it happily ever after.

Within days I found another tile dealer, this one in the New Haven area. When the young yard manager learned we were building a solar house, he immediately gave us preferential treatment. He patiently listened to our tale of woe about his unfortunate predecessor and assured us that he would try to make it up to us.

After choosing tiles for the bathroom floors, I came across a lovely unglazed quarry tile in a rich warm tone. Its matte finish was sand-flashed and its subtle shades ranged from a dark, earthy brown to a lively copper glow. It would be a perfect complement to the predominantly blue and green furnishings. Well-meaning friends predicted that tile flooring would be too hard and would cause all manner of ills—phlebitis, thrombosis, and what have you. They were certain it would be cold and noisy, especially when it echoed the sound of china or crystal shattering into a thousand pieces as it struck the tile. But C.B. and I were determined to have tile and we've never regretted our choice.

The design of the fireplace and hearth also had to be settled before we could put the specs out for bid. Again Don and I found ourselves at odds. We wanted a traditional masonry fireplace and Don wanted a metal firebox with an exposed chimney reaching to the ceiling. Don pointed out that conventional fireplaces could and did lose an extraordinary amount of heat up the flue, whereas a manufactured metal firebox could circulate heated air through ducts behind the firebox and recirculate the warmed air back into the room through open registers in the front of the fireplace. It would also be more economical. But we still preferred large brick fireplaces with expansive hearths, such as the one we had in our Stamford home. To us, a metal firebox hardly seemed a fitting substitute for the beauty and warmth of masonry. Besides, the fireplace would be an important architectural element in our solar home; it would become the focal point of the living room after dark or whenever outdoor visibility was limited.

To assuage us, Don encouraged C.B. and me to visit dealers in the area before making a final decision. We had to admit that some of the freestanding metal units were colorful and charming, but they were all much too small for our setting and hardly imposing enough to become a focal point for our large living room. Anyway, we had plenty of space; why did we have to skimp on the fireplace? Why couldn't we have one in brick or stone?

Don stood his ground, insisting that we'd save hundreds of dollars by using the manufactured firebox. Don won this round. We ended up with a commercially made, recirculating firebox. Later, with a generous facing of the copper-colored floor tile and a greatly extended hearth topped with the same material, the fireplace took on a new appearance blending well with the decor of the rest of the room. And I must confess the fireplace is probably more in keeping with the contemporary design of the house than a traditional fireplace would have been.

There were other equally critical areas where Don and I did find ourselves in complete harmony from the outset. We agreed that building in most of the furniture would make the house less cluttered and much more efficient. By using the same material for constructing all the cabinets and built-ins, we hoped to create a pleasing uniformity. Once again, Formica seemed to be the best option. It is durable and easily maintained, and it comes in a wide variety of colors and textures. And for this purpose it would be far superior to wood, which might warp, or to metal, which might corrode in our damp climate. Don suggested a dark, smooth-grained walnut, but I could only visualize this in the office of a bank president. Instead we chose a shade called honey-teak, a delicately grained and textured pattern in varying shades of brown with gold highlights.

When it came time to submit the plans and specs to builders, the detailed descriptions for the built-in sofa, bookcases, cabinets and other furnishings were conspicuously absent. At the time this appeared to me to be an oversight or a mistake on Don's part and it led to our mutual irritation and frustration. Long after the house was built, I realized this was not the case. I learned then that it is good architectural practice to keep the plans as simple as possible and to avoid too many details. This way, it is hoped, the builder will come in with a low "base bid." Another logical reason for keeping the initial plans simple is this: as construction progresses, it becomes much easier for both the owner and architect to visualize each section of the house and to make more appropriate choices and wiser decisions.

However, leaving too many details out of the plans and specs also has its disadvantages. The details left out until after the builder gets the bid are then added on to the base bid. Later on, this often ends up being very costly and, in some ways, you have only put off the inevitable.

The specs, for better or for worse, were now complete. All those choices and decisions we'd been making would soon materialize into truckloads, crates, and containers delivered to the site, somehow to be manipulated, applied, assembled, and, through the magic of construction, transformed into our solar house.

3

Difficulties Before the Building Contract

There is a fine line between comedy and tragedy, or so it is said. I think the placement of this line has a lot to do with time. In retrospect, some of the incidents in our solar adventure seem pretty funny. But when they actually happened, I don't recall having laughed. Such was definitely the case the day Don Watson presented us with a model of our solar home.

Early in June of 1973, Don called to invite us both to his office. When we arrived he greeted us with a little smile and a sly, droll expression. As he backed over to his drafting table, we could see he was carrying something behind him. With a deft maneuver and a backward lurch, he plunked down his hidden surprise.

The "surprise" looked strangely familiar yet totally strange. C.B. and I simultaneously gasped a quiet "Oh!" as it dawned on us that this odd-looking, squatty, three-dimensional rectangular figure with those unusual dark bumps on top was a scale model of our solar house! It was difficult and painful to conceal my disappointment. I couldn't believe that this weird model was what we'd been beating our brains out to create for the past six months —our beautiful dream house. It looked nothing like anything I'd imagined in my wildest fantasies; there were no sharply angled gables and no steeply pitched roof. It had such a raw and unfinished look! It reminded me, for some strange reason, of a newborn puppy as it first greets the world, trying to wriggle out of its fetal sac.

When I finally summoned enough courage to stop staring at the model, I looked over toward C.B. I couldn't believe it! There he was smiling and all relaxed, looking totally unconcerned. One thing was sure, neither he nor Don was sharing my sentiments; they didn't even seem to be aware of my startled consternation.

If those collectors had ever been in the plans, I certainly had missed them! I couldn't remember seeing them, and we certainly had never discussed them or exactly where they'd go. Even so, I should have known better. I must have been far too preoccupied with doors, cabinets, and tiles to have given the drawings of the elevations a much-deserved second thought.

I couldn't believe I could have been so naive as to think that all the talk about solar equipment wouldn't end up radically changing the outside of the house. I couldn't remember what I had imagined the solar units might look like or even if I had ever stopped to think about them. I guess I'd just assumed they would be largely invisible, like elevator terminals or air conditioning units on top of apartment buildings.

By this time, Don was fondling the model as if it were some adored toy. When he finally separated himself from it, he launched into a rather lengthy explanation of the solar aspects of the house.

The first row of collectors would consist of fourteen slanting units, each 6 feet high and 2 feet wide, stretching almost across the full 28-foot width of the south side of the house. The second and third rows of collectors would each have eleven units, making a total of thirty-six collectors with a combined area of 430 square feet. Allowing 10 to 15 feet of space between each row of collectors would mean covering almost the entire 1,000-square-foot roof. Completely dominating the top of the building, the collectors gave the house a flat, top-heavy look. It reminded me of a fortress from some planet in science fiction. The more I stared at the model, the more bizarre it looked!

Don was still droning on about the solar features, entirely oblivious of my distress. The collectors had been designed as an integral part of the roof so there would be less "architectural constraint," he said. They would be arranged in rows in sawtooth fashion with enough space between them to assure that they would not interfere with the collection of solar heat by casting shadows over each other.

As Don went on, I began to feel myself losing control. Who cares about the damned heat? Why did that have to be such a big deal? We'd be dead long before this country ever ran out of oil. What about what was happening in the here and now? How had my beautiful little seaside cottage turned into this freakish nightmare?

C.B. and Don never stopped chattering. After thirty years of marriage, couldn't C.B. see I was upset, seething, about to explode? Meanwhile, a rapid-fire dialogue was in progress inside my head. One part was saying, "So what if the house isn't the most beautiful in the world? What's the difference? How important are externals? Aren't the comfort and livability of a house far more important?" But the other side kept answering back, "That's a lot of nonsense! You bet appearances are important. How do you think you'll like looking at this every day for the rest of your life? Besides, if you're going to live in a weird house people are going to think you're weird too!" Why couldn't we have a *beautiful* solar house instead of an *ugly* one?

Luckily, neither Don nor C.B. could hear this disturbing mental discourse. But once I stopped repressing the idea that the model *did* look ugly, I began to feel better. I could learn to cope with ugliness. Couldn't you always learn to love your own child, no matter how ugly he or she was?

Now Don was grinning like a Cheshire cat and still purring about his handiwork. And C.B., probably exercising his inimitable tact and tolerance, had stopped talking about the model. They seemed to be discussing everything *but* the model.

But even as I sat there smoldering and miserable, another unhappy thought crossed my mind: what had become of all those exposed beams in the living? They were certainly not visible in the model. Some weeks ago Don had spent the better part of an hour patiently explaining the difference between authentic beams made of aged wood, and the common, commercially manufactured variety often made of plastic or some other synthetic material. Although these artificial beams were much cheaper and looked almost the same to me, Don had strongly advised us to stay with the real thing.

But where was the real thing? When I bluntly confronted Don about the beams, he blithely responded, "Oh, they're gone! There was just too much going on up there!"

Going on up *where*? On the ceiling? What in damnation was he talking about? I couldn't see anything "going on" in the ceiling of the model. Not even when you lifted off the top and looked inside of it. I had always wanted to have exposed beams and I had expected them from the beginning. This was just too much!

When I turned to C.B. for consolation, I found his attitude almost as maddening as Don's. He was sitting there looking so pleased with everything, joshing around with Don, ignoring me. After all those years together couldn't he tell I was about to scream? At this point I'd have gladly settled for some tawdry little condominium where I'd never have to worry about what was "going on" up on the ceiling! Who ever said building a house was fun?

But there wasn't much sense wasting any more time lamenting the departed beams, especially if C.B. didn't give a hoot whether we had them or not. Now the question was what to do about this queer model.

I kept trying to reassure myself that I could learn to love even this house once it was mine. Though the outside of the model wasn't at all what I had anticipated, the interior plan did seem to have almost everything I'd hoped for. Seeing the model in three dimensions was radically different from seeing the plans on a flat drawing. The high ceilings and the minimal use of walls combined to create a feeling of spaciousness and the pleasing illusion that one space naturally extended into another.

Basically, the house would be a split-level, with a half-story between each floor. There would be five levels in all, the basement, the garage, and three living levels. There would be no wasted space and there would be adequate privacy and separation between each level and each section. The house would be big enough for large-scale entertaining, but it wouldn't be so large as to make us feel uncomfortable when we were alone. Don had even taken the trouble to make stick figures scaled to the model.

Actually, Don had gone one step further and had taken color slides of the model set against a simulated beach background. When we viewed the house this way, dramatically silhouetted against a pale blue sky from some distance back, it looked quite remarkable. The beach location set off the bold lines of the house. It looked as if it belonged there and had always been there.

By the end of this conference, I was feeling less frantic and somewhat more kindly disposed toward the model and its creator. Don urged us to take it home over the weekend. He said this might be our last opportunity to make any fundamental changes before the plans went out to bid and the bulldozers gouged out that first shovelful of sand.

Before we left Don's office, I managed to muster a few lukewarm compliments about the model. On the way back to Stamford, C.B. said very little. Finally, I couldn't bear the silence any longer and blurted out, "How did you like the model?" While I strongly suspected what his answer would be, I wanted to hear it from him before I declared myself. He said enthusiastically, "It's great! Why? Didn't you like it?"

Just as I thought! C.B. has always been far more accepting and tolerant than I. "No," I said. "I really didn't like it very much. I thought it was very odd. Frankly, I think it's *ugly*!" C.B. reluctantly admitted that it was very unusual, especially the roofline. "But," he added, "we should be proud to show off that roof with those solar things for the whole world to see."

Here was C.B. calmly assessing and accepting this strange-looking house, while I was confused and unhappy, slowly working myself up into hysteria.

Knowing I'd feel better after I let off steam, C.B. gave me a loose rein. He's a wonderful listener and I had plenty to say about building, about solar heating, and about a certain solar architect! I ranted and he listened. After this catharsis I began to feel much relieved.

Some close friends dropped in over the weekend while we still had the model .We had learned to expect raised eyebrows, skepticism, and downright criticism whenever the subject of solar heating came up. A few of our friends had already warned us against being guinea pigs, or being "the first." Many had cynically remarked that if solar heat was so great, why hadn't anyone ever heard of it? And why hadn't people been using it long before this? Other friends came up with what we soon learned was the stock solar question: "But how do you heat the house when it isn't sunny?"

Yet, somehow, whenever we tried to communicate any of our recently acquired and sketchy solar knowledge, most people turned a deaf ear. Some expressed surprise or even astonishment that a hardheaded, successful businessman like C.B. would allow himself to be hoodwinked into such a wild experiment. Then there were those who didn't hesitate to tell us they felt Don and Ev were just using us to further their own careers at our expense. Literally no one gave us the kind of encouragement we needed and hoped to hear. None of our friends, acquaintances, or colleagues seemed to share our rapidly growing enthusiasm for solar heating.

Our weekend visitors were no different. Some of them actually seemed to recoil from the model. Then they'd force some innocuous comment like "How interesting!" or "it certainly is different!" Then they'd quickly change the subject, trying to hide their embarrassment.

On the following Monday morning, when it was time to return the model, I was somewhat amazed to find that I had actually grown rather fond of our "new baby." It no longer seemed alien and unfamiliar. And I had done a lot of thinking about it. After all, if no one were ever willing to surrender the tried, true and familiar, we'd never have anything new or different; we'd all be hung up with the same old status quo forever.

What if our solar house wasn't such an eye-catcher? We'd still have the last laugh on everyone when the country ran out of oil. Maybe then they'd realize that we weren't so foolish and that this solar heating wasn't such a bad idea after all.

At our next meeting with Don, we returned the model with our blessing. Now that the plans and specifications were ready and we had given our official approval to the model, it was time to look for a builder.

But before we could get started on this new phase a personal near-tragedy preempted our attention. For several weeks C.B. had been having some unusual physical complaints, although he wasn't in any pain. Neither of us thought it was anything serious until the doctor put him in the hospital for observation. When the tests and procedures dragged on for more than ten days, I became uneasy and set up an appointment to see C.B.'s specialist.

I was totally unprepared for the doctor's direct and rather brutal statement, given with no preliminaries: C.B. was suffering from a malignancy and had less than a year to live.

For a moment I was utterly speechless. When I didn't respond, the doctor asked if I had been listening. Of course I had been listening! I was just too numb to say anything.

When I visited C.B. at the hospital later that day, he already knew about the diagnosis but pretended to make a joke of the whole situation. I found it difficult to play along with this. Over the years I had come to take for granted C.B.'s broad shoulders, patient ear, and eternally optimistic outlook on life. It was impossible for me to contemplate this big, vigorous, and vibrant man reduced to the despair and devastation of a terminal illness.

In the early part of our marriage, when tragedy had struck our youngest child, C.B.'s bright and buoyant spirits had helped me face our loss and learn to live again. C.B. had savored every moment of living and, though in his sixties, had as much interest and enthusiasm as he had when we first met. I didn't dare let myself think what life without him might be like.

Within a few days, C.B. was discharged; he would be allowed to continue working as long as he felt well enough. He was given new and stronger medication and would have to remain under constant and close medical supervision.

It took awhile to regain my balance after this staggering blow. For days I didn't give the solar house a thought. When I finally did consider what effect C.B.'s illness would have on our building plans, I found myself in utter confusion.

What were we doing building a beach house that C.B. might never be able to enjoy? How could we spend all this money on a house when we might be needing every cent for medical care? On the other hand, how could we cancel the architect and engineer who had already invested so much of themselves in our solar house?

Since C.B. seemed to be taking things in stride, I hesitated to discuss anything relating to the future. After many sleepless nights, feeling the need to talk all this out, I confided in a close friend. Her immediate reaction was to say, "If you have any doubt, give up the idea of building, now, before it's too late." I found it difficult to subscribe to this line of reasoning, but I did give it some thought. Later I consulted another close friend. She pointed out how much building our beach home meant to C.B., how eagerly he was anticipating retirement and living on the water, and how much his hopes for the future were tied up with the house. I suddenly realized that I agreed with her thinking and that for however long C.B. might be able to enjoy the house, it would be worth building. She helped me recognize that living at Pilots Point, which C.B. loved so much, might prolong his life or even result in remission.

It was only when I heard her say the words that I knew these were precisely my sentiments. I felt renewed and reassured that we should go ahead with our plans to build.

Meanwhile, life seemed to be going on as usual, although nothing could really ever be the same with this sword of Damocles dangling over us. I tried to conceal my panic and anxiety, so that we could enjoy whatever time we had left together. I'm not sure how successful I was, because I did find myself becoming more and more obsessed with the thought of C.B.'s terrible illness.

I must have a very willful, rebellious streak. Whenever anyone hands down what seems to be an arbitrary or authoritative opinion, I automatically challenge it. When I was able to put distance between me and C.B.'s specialist at the hospital, I began to wonder about his diagnosis. I knew malignancies were difficult to track down without lengthy surgical procedures and that miraculous cures had been known to occur. Though still trying to accept the diagnosis and cope with the reality of C.B.'s illness, I decided to write my cousin Ed, an eminent pathologist practicing in the Midwest. Ed has always been like a big brother to me and had bailed us out of earlier medical crises. Although I really doubted that he could help us in this critical situation, I knew he would do his best. In any event, he would want to know about C.B.'s illness.

Within a few days, Ed called and asked me to send C.B.'s medical records along with the latest test findings. Ed called back a week or so later, and told us he had many reservations about the specialist's diagnosis. He hadn't found positive evidence of malignancy, although he was fairly certain C.B. did have a serious disease. While it might be incurable, it definitely could be treated and, in his opinion, was neither malignant nor terminal. Ed then recommended that we go to see an old colleague of his in New York for further evaluation and consultation.

We were afraid to get our hopes up too high. This was the first real encouragement we'd had through C.B.'s entire illness. Despite our efforts to stay low-key, our spirits soared and we could hardly wait for the appointment with the New York physician. It was remarkable if not miraculous for us when this kind and gentle doctor not only thoroughly agreed with Ed, but also said he was fairly sure he could treat C.B. and stabilize his condition. The news made us positively jubilant. We celebrated by having our first drink in weeks and found ourselves getting giddy and lightheaded. It was hard to believe our horrible nightmare was only an unfortunate fact of life, and one that we could learn to live with.

With this disturbing episode behind us, we could turn back to the drawing board. Now that our personal lives were again on an even keel and the architectural drawings were in order, we were ready to look for a builder. Even with the best architectural plans, without a competent builder to execute them, we would only court disaster.

Only the builder's workmanship and reliability could bring the house to satisfactory completion. And in our case the builder would need to have qualifications over and above the usual ones. He would have to carry out the architect's and engineer's instructions about installation of the solar elements. He would also be called upon to use new approaches and techniques and to work with unfamiliar materials and products. As the third member of our solar team, he would have to work closely and continuously with the architect and engineer and accept their constant direction and redirection. A recalcitrant builder could create unnecessary problems that might impede or impair construction. In other words, the builder could make or break our solar house. The more we thought about it, the more we realized there are hundreds of ways an unscrupulous builder can cut corners by using shoddy materials or incompetent labor. Choosing the right builder would be as important as choosing the right architect.

Don said he would be contacting some builders in the area and encouraged me to do the same. It was a relief to have an active role again and some specific responsibility.

On July 7, 1973, armed with the black-and-white plans and the sun-yellow specs, I once again set out in search of a builder. It was C.B.'s choice to continue to remain on the sidelines, as a one-man cheering section. He thought I was brave but foolish to try to land a builder singlehandedly, knowing so little about construction in general. My determination could probably best be described by that timeworn adage "Ignorance is bliss."

My first call was to a local builder who expressed great interest in our project and asked me to drop off the plans. Maybe this wasn't going to be so difficult after all. When I didn't hear from him for several days, I began leaving urgent messages at his office. A week passed but the messages remained unanswered. It began to look as if I was getting the runaround, so I decided to retrieve the plans. When I did, he returned them to me with pleasure. "This isn't a normal house," he jeered as he tossed the plans roughly in my direction. "Whoever heard of a silicone roof and solar heat?"

Somewhat beaten, but unbowed, I tracked down another well-known builder and gave him a copy of the plans and specs. Two or three days passed before I reached him again. When I did, he seemed uncomfortable and evasive. First, he said he hadn't had enough time to study the plans thoroughly. Then he said he was pretty busy right now. Finally he blurted out, "Where in the world did you ever get this idea of building a solar house? I never even heard of one and I sure wouldn't want anything to do with building one!" Another builder to be scratched.

I tried a third firm which assured me that they wouldn't be able to consider a contract for less than $80,000—twice what we planned to spend!

Just as I was beginning to despair of ever finding a builder, Don called. He wanted to know how soon we could meet with an Essex builder named Richard Riggio. An appointment was set up for August 15. Ev would join us so he could explain some of the more technical points about solar heating to Mr. Riggio.

This initial meeting got off to a less than propitious start. While the four of us waited almost an hour at another meeting place, Mr. Riggio paced up and down at the building site. When we finally joined forces, Riggio was good-humored about the confusion, jokingly saying that he hoped we'd develop a little more "togetherness" once he started building our house.

Riggio, or Dick as we were soon to know him, looked more like an outdoor sportsman than a general contractor. Only his mud-caked shoes were a dead giveaway of his profession. He has a warm, outgoing, gentle manner and, although middle-aged, looks years younger because of his ruddy complexion and thick wavy hair. His enthusiasm for building a solar house was immediately apparent and he made no secret of his eagerness to be awarded our contract. Don was lavish in his praise of Dick's reputation as one of the most outstanding builders in the area, if not in the state. There was little doubt where Don's vote would be going.

Don kept referring all our questions to Dick, who fielded them exceedingly well. There was no doubt that he was not only a very experienced builder but also a very intelligent man. Dick didn't seem to anticipate any special problems because of the house's solar features. He was currently finishing up another job but was sure he could start building our house by mid-September. At last everything was falling into place.

The meeting ended on a hopeful, almost joyous note. Before Dick left, he urged us to visit one of his houses to see some of his workmanship firsthand.

We set out eagerly the very next afternoon and found one of Riggio's homes nestled in the hills overlooking the Connecticut River. A large cantilevered redwood deck with enormous built-in planters ran around three sides of the building. It was long past working hours, so there was no one around, and we slipped into the house through an unlatched sliding glass door.

The living room was extra large with dramatic skylights dotting its cathedral ceiling. A very high window-wall framed the spectacular view down into the valley, across the river, and beyond to a rolling ridge. The paneling and built-ins were exquisitely finished and were fine examples of Dick Riggio's special talents as a craftsman. We began to feel very lucky that Don had found such a gifted person to build our home.

Two weeks later when we saw Dick again, we remarked that the house he had built along the river was "very beautiful." He thanked us and quickly countered, "We think your house is going to be very beautiful too!"

How refreshing! What a far cry from the criticism, skepticism, and outright scorn we'd been encountering. Our confidence in Dick was rising high and fast, and basically that opinion has never changed.

Within two days Dick came up with a bid of $60,500. While this wasn't as bad as the $80,000 estimate, it was still much more than we intended to spend.

C.B. and I needed some time to consider just where we were and just where we were going. By this time, we had paid out some rather substantial sums. The land had cost $20,000; the legal fees for purchasing the land, $1,000; the white cedar siding, $2,500. That much alone added up to $23,500. Besides this, we had just made our third payment on Don's architectural fee. We had already paid him $2,100, with only $700 outstanding. That final payment would have to be made soon after construction started. In addition, the cost of neither the auxiliary heating system nor the solar heating system had been included in the specs. Since Ev had estimated another $7,000 for them, the grand total of expenses amounted to a cool $33,600 so far—and this before even one grain of sand had been disturbed.

These figures seemed somewhat mind-boggling to me. We'd started with a $40,000 total budget and Dick Riggio's bid for $60,500 alone was already over a third higher than this.

Just the thought of spending such large sums of money made me very uneasy. We had both worked hard to save a respectable nest egg for our retirement. But at the rate we were going, we'd probably have to blow quite a bit of it on our new home. We had always lived comfortably and well and hoped to maintain our present standard of living when we retired. I certainly did *not* want to tie up everything in our new house.

But C.B. didn't share my views. More accustomed to dealing with large sums in business, he had an entirely different perspective on the whole matter. He didn't think Dick Riggio's bid was at all out of line and he felt strongly that we should accept the bid right away. C.B. tried to explain to me at some length that our original budget of $40,000 had never been realistic. Also, since we'd first come up with that figure, the cost of labor and materials had jumped tremendously. In 1972, building costs were calculated at roughly $25 per square foot. Only a year later, $30 per square foot was a much more realistic figure.

Then, too, the final plan of our 2,000-square-foot solar house was a far cry from the simple little cottage we'd originally contemplated. C.B. said he never would have been satisfied cooped up in a few small rooms in some tiny house. All three of us are . . . big! C.B. is six feet two and almost two hundred pounds; Steve is even taller, but slimmer. I'm about five feet four and my weight, though a carefully guarded secret, is not to be dismissed lightly.

Besides, each of us has many interests and hobbies. C.B. is into computers, boating, and fishing, and all the paraphernalia that goes with them. I have collections of paints, shells, and sewing materials, and a fair-sized professional library on psychiatric subjects. In addition to hundreds of books, Steve has a huge assortment of athletic gear—outfits and equipment for soccer, rugby, tennis, and fencing, skis for land and sea, as well as extensive camping gear. And as a family we've shared our interests in stereos, sleds, toboggans, ice skates, golf clubs, surfboards, and bicycles. C.B. was right! We probably couldn't have squeezed ourselves and all our accoutrements into a smaller house; nor would it have made any sense to put up a very inexpensive home on such a costly piece of property. Nevertheless, I still had reservations and decided to meet with Don once more to see if there were any ways to reduce our building expenses.

Don was sympathetic to our concerns about the budget and to the fact that we really had received no competitive bids. He did remind us that despite our combined efforts we had only been able to produce one other bid, which had been almost $20,000 higher than Dick Riggio's. Don was pleased with everything he'd learned about Riggio and was thoroughly convinced he was the right man to build our solar house. Assuming we could get lower bids, Don explained, accepting the lowest bidder isn't always the wisest policy. We had to balance the lowness of the bid with the reputation and competence of the builder—or we might end up with a poor builder and a badly constructed house.

Don did offer several suggestions for bringing down building costs. One way would be to do all our own contracting. This meant we would hire our own carpenters, plumbers, electricians, and painters without benefit of the overall services and supervision of a general contractor like Dick Riggio. By thus eliminating the contractor's usual profit of 10 percent over and above costs, we might be able to save $6,050. Don was fairly certain we could trim the budget further by using a few of the students he knew to do some of the carpentry and other manual construction work. He thought they would be competent and willing to work for less than most union laborers.

On the surface, this seemed like a wonderful and relatively painless way to reduce our expenses by $6,000 to $8,000. But when C.B. and I went into this in more depth, we realized that for us to attempt this could only lead to disaster. We were babes in the woods as far as any knowledge of construction went. And even if we did have the necessary know-how, with both of us working full-time at jobs many miles away, we couldn't possibly be on hand to supervise and deal with a half-dozen different subcontractors. We decided not to pursue this alternative.

Don also suggested that we could postpone putting out any more bids until the middle of winter, some five or six months hence. By the beginning of January, he said, the building trades would be slowing down and we might find builders a lot hungrier and more anxious to put in bids, even bids for a solar house.

But the idea of postponing construction for several months was not very appealing either. We had already spent a full year planning and were becoming very impatient to get on with it. Also, we were afraid that by January the cost of materials and labor might be even higher and would therefore cancel out any possible savings we might make.

Another money-saving alternative was to eliminate everything but the essentials of the house. But on reviewing the plans with Don, we found few things that we could—or wanted to—eliminate. We also feared that in cutting back we might destroy the essence of the house.

There was clearly one other obvious way of reducing costs: we could give up the whole idea of solar heating. This would be another relatively painless way of reducing our budget by thousands of dollars.

It wasn't until this juncture, when we were forced to reexamine our priorities, that C.B. and I realized how deep and firm our conviction to build a solar house had become. We were now fully committed. The solar features of the house had become an indispensable and intrinsic part of our thinking and we were not about to give them up.

Energy shortages had grown worse during the months when we'd been planning the house. Solar heating was looking more practical and economical every day. While the initial cost of the installation would be considerable, we were both thoroughly convinced we would save a substantial amount on fuel in the years to come. So, after much soul- and budget-searching, we decided to accept Dick Riggio's bid.

The building contract was signed on August 18, 1973, despite my lingering misgivings. We agreed to pay Mr. Richard Riggio of Melody Lane, Essex, Connecticut, $60,500 to build our solar house.

Our building contract was a standard AIA "small construction contract." The total cost of construction was fixed and spelled out in the contract. It included a "bonus agreement" as an added incentive to the builder, providing that whatever amount the builder was able to save under the "stipulated sum" would be considered a "bonus" to be divided equally between the builder and us, the prospective owners.

We would pay Dick Riggio for his services as our general contractor in "progressive monthly payments" as the work was completed, providing the bills had been approved by Don.

Construction was to begin "on or before" September 15, 1973, "weather permitting," and to be completed, "weather permitting," on or before March 14, 1974. Although most private homes can be built in three months, our solar experts were allowing an extra three months.

Construction would be in three broad phases. During the first six weeks the foundations, side walls, doors, windows, and roof were to be completed, making the house weathertight. The solar collectors and the large solar storage tank were also to be installed during this time.

Once the house was closed in, the plumbers, electricians, and solar engineer would be able to rough in their work, from about the seventh to the eleventh week. In the third and final phase, from the twelfth to the sixteenth week, the built-ins, cabinets, staining, painting, and all other finishing touches were to be completed. With this timetable, construction might take only four months and we might be able to move into our solar house by mid-January!

4

Permits and Problems

Most modern builders have never heard of designing with nature —a critical concept in solar home-building. Of primary importance to our home would be many natural factors often ignored even today when conventional homes are built. Traditionally, private houses have been oriented toward the street or road on which they're located, usually directly facing the thoroughfare. Interior layouts are most often arranged with the major living areas—kitchen, living, dining, and family rooms—looking out to the street.

Most houses are situated with little regard for maximum exposure to the sun and minimal exposure to the cold north winds. Important climatic factors—such as the direction from which most storms come—are usually lost among other considerations, principally those relating to the builder's budget. (It is true, however, that most builders are willing to take into account one natural factor: the view. Even the most mercenary builders are usually willing to position a house so that any special scenery can be enjoyed.)

I've noticed that while many builders and real estate agents are aware of climatic factors, they tend to consider them only when it suits their purpose. They might, for example, point out that a master bedroom has ideal southwest exposure or that a kitchen will be bright and sunny all day. But they are far less likely to call a client's attention to the tall pines completely shading the house, or the fact that most of the doors and windows are on the cold north side of the house—factors that will inevitably result in higher heating bills. Nor are they apt to point out that an abundance of west-facing windows will mean greater exposure to the hot afternoon sun in the summer months.

These natural factors are probably given somewhat more attention in architect-built homes. Architects are far more likely to consider the advantages of different exposures and positions before situating a house on a piece of property. Surprisingly enough, however, most architects do not observe these basic principles with any consistency.

While climatic factors are important in positioning any house, they are indispensable considerations in planning a solar house. Solar efficiency can best be achieved by taking full advantage of all natural benefits.

Because of the small size of our lot and the relatively large size of the house we planned to build, there would be almost no room for adjusting our house in any direction. There was, however, one important feature of our land which made it a "natural" site for a solar house: the beach side of our property faced due south.

When Don and I worked out the plans, we had considered the function of each room in relation to the natural benefits its location and exposure would bring. The main living areas were grouped on the south side of the house, allowing them to collect the greatest possible amount of solar heat passively. To avoid excessive heat in summer, openings on the west side of the house had been carefully planned and limited. The main entrance on the north or street side would be recessed to provide extra protection from high winds and wet weather. And the placement and number of north-facing windows would be carefully assessed, as it was here that we would risk losing the most heat.

There were other ways in which our solar house would gain advantages from nature's bounty. Great care had been taken to insure excellent daylighting and cross-ventilation to reduce the need for artificial light and for mechanical heating or cooling wherever possible—providing this arrangement did not interfere or detract from our basic comfort or aesthetic goals.

The house itself, in order to comply with local zoning regulations, would have to be centered on our roughly to-by-90-foot lot. (Actually our eastern boundary is 102 feet.) Since the house we planned would measure 54 feet in depth, this would leave a total of only 44 feet of land on the north and south sides of the house. From this remaining 44 feet we would need to deduct front and rear setbacks of 20 feet each. With even my limited mathematical abilities, it was clear that we would have at most 4 feet within which we could move the house in a northerly or southerly direction. (While north and south are easy enough to define, C.B. and I still find ourselves poles apart on our definitions of what "front" means. I have always considered the street side of our property as the front; but C.B. has just as consistently thought of the beach side of the house as the front. To this day, if we agree to meet "outside, in front," C.B. unhesitatingly heads for the beach while I instinctively rush toward the street.)

Other local restrictions would need to be considered before building could start. Since our total land area amounted to less than 6,000 square feet, there would be no way we could possibly conform to an ordinance requiring a building lot to have a minimum of 10,000 square feet. Another restriction specified that a building could not cover more than 25 percent of the total land area. This meant our 6,000-square-foot lot could have a house no larger than 1,470 square feet, or roughly 500 square feet smaller than the house we were planning. These seemed like impossible limitations for our home and we were very relieved to learn that they did not apply to us. All the land on the Point had been developed before the existing zoning regulations went into effect. Our small lot, therefore, was simply designated as "nonconforming," as opposed to being in violation of the law.

Our house plans would also need to be approved by our local property owners' association, which set minimum standards for the size and quality of the homes to be built and raised funds for maintaining the roads and beaches. Getting the association's approval of our plans proved to be quite routine as the incumbent president quickly gave us his official blessing.

Another matter for official approval was our septic system. Because our lot was so small in comparison with the size of the house we hoped to build, it would be difficult to work out an adequate arrangement for sewage disposal. We just managed to squeak by on this one. New legislation requiring much larger reserve areas for septic systems had recently been passed but, luckily for us, would not go into effect until January 1, 1974, three and a half months *after* our construction was due to get under way. The new law would require reserve leaching fields equal in size to the septic system itself. It would have been virtually impossible for any contractor to comply with the new regulations on our limited land area. As it is, our sewage disposal system resembles an intricate maze.

By far the most difficult restriction to observe was the legal limit on height. Because of the marina, the Point is classified as a "designed boating district" with a height limitation of 20 feet on all buildings. While this kind of regulation might not have interfered with the design of a traditional home, it could present serious problems in building our multilevel contemporary house and especially in regard to its solar collectors.

Although we would have just three stories of living spaces, our house would actually be built on five different levels: the basement at below-ground level, the garage at street level, and the three floors of living space above both of these. Basically, it was a split-level plan. With only four feet between each of the floors, the house would rise two and one-half stories above the street. Whether we could stay within the designated 20-foot limit would actually depend on the height of our interior ceilings, usually 8 feet. With two and one-half stories, 8 feet to each story, our house would be exactly 20 feet high.

But we weren't in the clear yet. If we added another six feet for the collectors we would certainly find ourselves out in left field. Don presented our plan to the building inspector and we waited anxiously. Since the solar equipment had put several local builders into a tailspin, we anticipated that the building inspector and the zoning board members might have reservations about our solar features. There was a good chance that none of them had ever heard of a solar collector and even a better chance that they had never seen one and hadn't the foggiest notion of what a collector looked like. Unless our town fathers were progressive and generous enough to consider our solar panels as exceptions to the height limitations, we would be sunk. If they did decide on a strict, narrow interpretation of the present law, we could always petition the Zoning Board of Appeals. But we'd have no reason to believe that this board would look any more favorably on our solar equipment.

Happily for us, our little unsophisticated town proved itself more forward-looking and liberal than many larger communities have since shown themselves to be. These country gentlemen quickly decided that the local height regulations did not apply to our solar panels. They chose to interpret the local statute to include the solar panels in the same category as "appurtenances" such as chimneys, spires, towers, penthouses, and tanks—all of which had already been designated as exceptions to the town's height limits. We were lucky to receive this interpretation. Today we still hear of solar builders who are held to a narrow interpretation of local building restrictions, sometimes making it impossible to put up a solar building at all.

So there we were, in the late summer of 1973, at last in the clear with our local officials and ready to build. Although there were only a few puny red-tagged stakes in the ground, we could look back over the past few months and realize how much progress we had made. It was now eight months since we had first heard Don's deep, resonant voice announcing his interest in designing our beach house. And for the past six months we had been learning a great deal about the wonderful ways of solar energy.

Not long before, we had come across a long article in one of the less technical magazines which answered a nagging question of long standing; namely, what had become of those early solar homes in Massachusetts? We learned that the MIT houses of the thirties and forties had been built primarily as engineering experiments, rather than as homes to be permanently occupied. After serving their basic experimental purposes, all had ultimately been abandoned. One of them was destroyed to make room for university expansion and another was divested of its solar equipment when the occupants got tired of maintaining it. A third house, rumor has it, was deserted when neighbors objected to the long lines of visitors to whom the owner had sold tickets of admission.

We also found out about another solar home which had been built around the same time in Dover, Massachusetts, some miles from MIT. This house was largely the work of Dr. Maria Telkes, a solar pioneer who is still active in solar experimentation today. The house Dr. Telkes helped design was said to be the first solar home to use a chemical known as Glaubers salt for storing solar energy. It was reported that this house had also been plagued by a continuous stream of visitors and, as far as we could learn, was no longer being used either.

Understanding more about these early solar homes proved one important thing to us; namely, solar energy *had* been used earlier for home heating with at least some measure of success. This kind of information was helpful to us in another way; it gave us fair warning of what could happen if we didn't take some preventive measures to control the number of visitors who might descend on us once we moved into the house. Without screening and selection, visitors could become an annoyance or even a serious problem.

The magazine article also gave a logical explanation of why the idea of solar heating had not taken hold in the thirties: in those days, fuels were so cheap and abundant there was no need to seek out alternative sources of energy.

But how different the situation is now, almost forty years later. Coal, oil, and gas have become scarce and expensive. And it looks as if we might exhaust our reserves of fossil fuels within the foreseeable future. Electricity is proving to be a very wasteful and expensive way of providing domestic heat. And the awesome hazards of nuclear power are still surfacing. It is, therefore, far more timely to experiment with solar heating today than it was in those earlier years.

While we struggled to get our solar house off the ground, the American people as a whole were struggling with the energy shortages. The situation was rapidly developing into a full-blown national crisis. The Middle East was shipping us less and charging us more for their precious oil. Suddenly everyone was becoming aware that few products or services would be unaffected by the sudden sharp rise in the price of oil. Everything was costing more: transportation and manufacturing of every kind and services of every description—from medical care to dry cleaning. People were finally becoming concerned with energy conservation. There was talk of rationing, wider use of car pools and mass transportation, reduction of speed limits, and, of course, lowering of thermostats. There was a renewed interest in insulation and its effectiveness in reducing heating costs.

For the first time, most of us were being forced to recognize our extravagant and wasteful habits of energy consumption. It made us uncomfortable to realize that we were using more energy in a week than almost half the people in the rest of the world used in a year!

But, despite all national efforts to conserve fuels, we were actually importing more oil than ever before. Our enormous and growing dependence on expensive oil from the Middle East was having serious and widespread effects on our economy. Because of the millions we were spending on foreign oil, our balance of trade was out of kilter. We were importing more goods than we were exporting, meaning there was more money going out of the country than coming into it, a condition known as an unfavorable balance of trade. This, along with rising inflation, was making our dollar worth less each day.

At this point no one seemed ready to face up to these cruel facts of life. Everyone seemed to be looking for a scapegoat. It was easier to blame something or someone else for the growing shortage of oil and its ever-increasing cost than it was to cut back on oil and look for new sources of energy. President Nixon soon became a prime target. He was already in so much trouble it didn't seem to matter very much if he was blamed for one more outrage. People complained he had been so busy trying to cover up his part in the Watergate affair he hadn't had time to develop a national energy program. Others seized the opportunity to place responsibility for our energy problems squarely on the shoulders of the Organization of Petroleum Exporting Countries. Weren't they blackmailing us by drastically reducing their exports of oil to us while simultaneously raising the price? And it looked as if the Arabs were going to continue their squeeze until we changed our policies toward Israel. Others accused our own giant oil companies of deliberately creating shortages in order to push up the price of oil and line their own pockets with excess profits.

While the American people continued to look for scapegoats, the government began taking a belated look at the possibilities for developing alternative fuels. Nuclear power was one relatively new form of energy that seemed to hold the greatest hope for the future. By 1973, more than fifty reactors were in use or under construction in various parts of the country. Yet even as we turned to nuclear power, critical problems were becoming evident—problems of expense, time, and safety. The cost of building reactors was growing exorbitant and construction itself required huge amounts of energy. Furthermore, it usually took ten years or more before a nuclear plant could become operational. Unforeseen delays and the advanced nature of the technology often meant that a plant would cost twice what had been originally planned. And who but the hard-pressed consumer would be asked to pick up the tab?

Accidents and shutdowns for reasons of safety and repair were becoming more common. Either there had been fewer accidents in the past or those that had occurred had been a carefully guarded secret kept from the public. One of the most serious hazards of nuclear power is related to radioactivity emission levels. What is a safe level of radiation? The government said one thing and the environmentalists said something entirely different. Because of radioactivity, transporting and disposing of nuclear wastes—which could remain radioactive for 240,000 years—were a growing menace to all of us. At this writing, a safe way to store nuclear waste (known as plutonium) is yet to be found. Nuclear accidents and the careless and hasty handling of plutonium could result in irreversible genetic damage to future generations.

Despite these very real concerns, in 1973 our government was spending 80 percent of all research funds earmarked for energy on nuclear power research! The remaining 20 percent had to be divided among all the other kinds of energy combined.

In competition for at least a tiny share of these funds was a "new," little-publicized alternative to fuel: solar energy. Proponents of this undeveloped but ageless form of power were slowly and quietly making their presence felt. A growing number of engineers, scientists, and research fellows in some of our leading universities were continuing to study and experiment with many applications of this greatest and ultimate source of all power: the sun.

When you stop to think about it, isn't it strange that scientists were only beginning to find such ways of using a resource which had been around since the beginning of time and which had influenced every moment of our existence? Without the light and heat of the sun, there could never have been any life on this planet in the first place. It is precisely and entirely because the sun has always been able to beam its fiery energy down to us through 93 million miles of space that the conditions for life exist. It is the sun's chemical actions and reactions which have always created and supported all forms of plant and animal life.

C.B. and I, with our rapidly expanding interest in solar energy, found ourselves searching for even tiny pieces of information about the sun's uses. We discovered some fascinating legends about solar energy and some of the ways it was used in ancient times. There is the classic story of how Archimedes used the power of the sun for military advantage. When he saw the Roman fleet approaching, he ordered his Greek warriors to focus the sun on their shiny shields and reflect its strong rays onto the enemy's sails, thus setting them on fire and repelling the invasion.

We came across more recent examples of how the sun had been harnessed by man to help him in his everyday life. We learned that there had been enormous solar devices operating in Chile for purifying water as long as a hundred years ago. And by the 1960s there were millions of solar water heaters in everyday use in Japan, and close to a hundred thousand in Israel. We were also surprised to find that for many years Australians had been successfully using solar energy to dry their crops and to heat and distill water.

Over twenty years ago, back in 1954, solar scientists and engineers from all over the world gathered at the University of Arizona to form what was later to become the International Society of Solar Energy. Since that time, the society has continued to spearhead the worldwide solar energy movement by sponsoring international conferences and symposia in many different countries, making it possible for scientists from almost every country in the world to exchange ideas and share results of their studies and experiments. The society's technical journal has become one of the leading resources for solar development in all parts of the world.

Even though the U.S. has been slow to recognize and put to use the sun's enormous potential, there have been a number of government projects and some federal funding for university research. In 1972 a solar energy panel was sponsored by the

National Science Foundation (NSF) and the National Aeronautics and Space Agency (NASA). The panel came up with some rather revolutionary and startling observations. Its members decided that solar energy could make a major contribution toward the solution of our energy problems and could significantly decrease our unhealthy and dangerous dependence on foreign countries for a large part of our energy needs. They recommended that the government spend $200 million a year over the next fifteen years (a total of $3 billion) to further the research and development of solar energy.

The following year, 1973 (when our solar house was already under construction), Congress appropriated $4 million—less than one-fiftieth of what the panel had recommended the government spend that year. Small as this sum was in relation to what was so desperately needed, it still represented a more than 100 percent increase over the previous year, when the allocation for solar energy was only $1.6 million.

Despite this lack of meaningful federal support, our major universities have been conducting solar research and experimentation for nearly two decades. By the seventies, the universities of Arizona, California, Colorado, Delaware, Florida, Pennsylvania, and Wisconsin, among others, were studying and experimenting with solar heaters, solar engines, solar cars, solar signals, solar watches, and solar houses.

Despite this rapidly growing interest, most people were still looking at us in utter disbelief when we told them we were building a solar-heated house. Usually, they were too uninformed and embarrassed or too confused and dismayed to venture any questions, and they would drop the subject quickly. Many of our close friends—and other people whom we admired and respected—continued to be skeptical and critical of our plans. They never lost an opportunity to continue their warnings that we were being "taken" by both our architect and our engineer. Later, some tacked the builder's name onto their list of villains who, they insisted, were exploiting us with their solar theories.

By now these dire warnings served only to reinforce our eagerness to get on with the building. At least that was one way we might someday prove how very wrong these doubting

Thomases had been. We could scarcely wait until September 15, the day when the ground on Lot 48 of Pilots Point would be broken.

5

Construction Gets Under Way

September 15, 1973, that magic momentous date when construction would begin, finally came . . . and went. Nothing happened. I made this startling discovery on September 19 after racing up to Pilots Point, breathless with anticipation, only to find Lot 48 as virginal and undisturbed as I'd left it weeks earlier.

What went wrong? During the weeks of rushing around to find a builder and getting all the local officials to approve our plans, I'd vowed to sit back and relax once the building got started. What did I know about construction anyway? I would sit quietly by and watch others work to bring our dream house to fruition.

Pure fantasy! The reality was that on September 25—ten days after the starting date stipulated in the contract—not a single grain of sand had been lifted or even moved around on our little seaside lot. C.B. and I looked forlornly at the pristine site, then at each other. Just what, we wondered, was going on? We called Don immediately but couldn't reach him. He'd be back quite late, his wife told us.

It was long past midnight when the phone finally rang. C.B. was asleep but fortunately I'd buried the phone under several layers of blankets to muffle its ring. (Unfortunately, I have never been able to perfect an equally effective technique for muffling my voice, especially when I am distraught and it is late at night.)

I could barely hear Don; he sounded faraway and exhausted. I felt a twinge of guilt for having insisted that he call me back no matter what time he got in. But didn't someone have to take responsibility for what was going on (or *not* going on)?

I angrily reported our discovery that there was no sign of activity at the site although the starting date for construction had passed. The builder had *already* violated the contract! After a long silence came Don's slow and weary reply, "Don't worry about the starting date in the contract . . . it doesn't mean a thing!" How infuriating! But the hour was hardly suitable for a long-distance debate and I hung up quickly.

This was great! If the contract didn't mean a thing, why had our lawyer scrutinized all that fine print? And why had we paid her a small fortune for doing so? And why, if the contract weren't worth the paper it was written on, had we bothered to have a contract at all?

The next morning I was still feeling angry and miserable. My first impulse was to call the builder, Dick Riggio, and ask him straight out why construction hadn't begun on time. But I soon remembered why I dared not do this: Don had already told us there was to be no communication with any of the men on the job. This would of course include the builder himself. I felt crushed and powerless.

The reason for Don's directive was a mystery to us at the time; in fact, it wasn't until long after the house was finished that we came to understand the real purpose for this ban on communication.

If the owner gives verbal instructions to workers, they might easily be taken as an authorization to change the plans their boss (the builder) gave them to follow. Also, a negative comment from the owner on some aspect of the work could not only confuse the worker but might divert him from his original task. If the project were to proceed in an orderly fashion, it was reasonable that only one person could be in command. (As construction progressed, we gradually got the courage to say a "Good morning" or a whispered "God bless you," if one of the workmen sneezed. As the weeks went by, it became all but impossible to clear each and every communication with Don and before long we abandoned the prohibition altogether.)

Happily, it turned out there was no need to contact the builder at this time. When we returned to the beach the following weekend, we were thrilled to find a plucky little yellow bulldozer busily gouging out huge chunks of sand, carefully piling them into one high heap. At last! We were on our way!

But as we watched this wonderful little scene, we began to realize that as fast as the bulldozer was scooping up the sand, a pint-sized dump truck was loading it, carting it away, and dumping it into the marina!

Don and I had always agreed to save every possible grain of sand, a precious commodity even at the beach. C.B. and I had heartily approved Don's suggestion that all the excavated sand be pushed back against the foundations as an extra bulwark against high tides and flooding. My immediate impulse was to dash over to the dump truck operator and demand that he stop this wasteful activity immediately. But again I remembered the restriction on communication, and instead I dashed to a phone and told Don all about the problem.

He must have taken care of things very promptly, because when we returned to the shore the following week, there was a huge mountain of sand on the beach. What was even more satisfying, the large wooden forms, into which the concrete foundations were soon to be poured, were already in place. The broad outlines of the basement were now plainly visible and we could see the shape of things to come. For the first time we could tell exactly how the house would be positioned and how much of the lot the house would cover.

Progress during these initial weeks was astounding. Construction seemed to be racing ahead. Before long the steel beams and Lally columns were in place in the garage and basement. By mid-November the first-level subflooring was completed; the following weekend the studs began to rise. At the end of the month the framing and flooring for the whole house were roughed in. At this rate the house would be completed long ahead of schedule.

It was exciting and rewarding to see the house take shape. We had spent so many months working on the drawings, we'd almost forgotten that they would eventually be transformed into concrete, glass, steel, and lumber.

Once the skeletal framing was finished, I found myself spending long weekend hours at the house. Just walking through the unfinished walls from one empty space to another, trying to visualize how everything was going to look, was a fascinating game. C.B. was beginning to tease me about spending so much time there. He wondered what I could be doing there and was very amused when I told him I often swept up sawdust and wood shavings and made a large collection of discarded nails.

I was beginning to feel a strong attachment for this empty shell that would someday be our home. For some reason I seemed to have a growing desire to protect it until it was big and sturdy enough to stand on its own. These quiet, solitary interludes were refreshing and seemed to restore my perspective. And it was satisfying to know that all the time we'd spent planning and running around had not been wasted and that all our hopes and efforts were finally beginning to materialize.

During these first few months of construction, we gradually became sensitized to the sights and sounds of the shore. We were fascinated by the screeching gulls swooping endlessly down and around us. We were getting accustomed to the sounds of power boats as they gunned their motors and sped off toward their favorite fishing grounds.

Sometimes the fog along the coast was so thick that the house was totally lost in a shroud of rolling mists. Often this heavy haze would hang over us for hours; then at last one lonely sunbeam would penetrate, reflecting millions of dewy particles floating in its path. Soon other rays would push their way through until the clouds would lift, quietly and mysteriously disappearing. Then the beach and sky would gleam and sparkle as the mighty sun broke through the heavens.

Even on dreary days there is much activity on the waterfront. Sometimes the wildlife preempts the scene from the boaters and fishermen. One late October day we became aware of an enormous flock of long-necked Canadian geese gracefully riding the shallows. We watched them feast greedily on the rich morsels left by the outgoing tide. After several hours, we suddenly heard a great squawking and cackling as the leaders rounded up these magnificent creatures. Within minutes, first singly and then in ever larger groups, they slowly rose into the air and dropped into perfect formation to continue their long journey south.

Fall sometimes brought unpleasant sounds too, like the sharp bang of the hunter's shotgun as he aimed at friendly little harbor ducks who had foolishly ventured out toward the islands. It was sad to think of their unhappy fate after we'd fed them in the harbor all summer and they'd learned to trust us so completely.

By the middle of November we had little time to savor the shoreline sights. We had resumed our conferences with Don and there was still much to be done to complete the plans. The builder would soon need the exact measurements and design of every cabinet, shelf, and built-in in order to carry out every last detail of the architect's design.

We were getting impatient to get on with these final drawings. But each time we asked Don about working out specific items, his stock refrain was "There's plenty of time!" But time was running out and we felt there was none to spare. Before long the carpenters, plumbers, and electricians would be wanting to know whether we'd prefer the kitchen sink centered under the window or a few feet closer to the stove; the size and style of the vanity in the downstairs guest bathroom; the height, depth, and number of cabinets and exactly where each one was to be installed; the size and style of the medicine cabinet for the master bathroom; how high the base for the sofa should be and how far it should extend into the living room; the exact dimensions of the fireplace opening, and its surround, and where, how high, and how deep the bookshelves were to be. The plans and specs would have to describe all this and much more. Otherwise, Dick Riggio would have no way of carrying out the architectural plans.

Here again my ignorance—and the lack of sufficient communication with Don—came into play. At the time I had no way of knowing it was good architectural practice to delay many of these decisions until construction was well under way, and you could actually see the completed areas—and thus be in a better position to make final choices.

One example of this process was the widening of the main deck. Although the deck seemed adequate on paper, once the shell of the house was up, we could see just how skimpy a 12-foot-wide deck would look in relation to the rest of the house. When we discussed this with Don, he was amenable to widening the deck to 15 feet. The additional 3 feet not only made the deck more in keeping with the rest of the house but also gave us a total of 570 square feet of outdoor living space.

Don had put off the final details for the fireplace until the firebox arrived. When we reintroduced this subject, Don agreed that it would be pleasing and practical to face both the fireplace and the large, raised hearth with quarry tile. In this way we would tie the living room in with the tile flooring in the kitchen and hall. The extra 24 feet of raised hearth would give us a charming and convenient place for plants, television, stereo, and just plain bric-a-brac.

Don had gone along with my idea of building in our large sofa or banquette, as these built-in couches are usually known. The banquette, designed in a modified "L" shape, would fit along the inside wall of the living room in front of the built-in bookcases. It would extend some 8 feet outward from the corner in both directions. This additional reserve of seating space made it possible for us to achieve another goal; namely, to keep the living room open and uncluttered. Since the banquette would parallel the center entrance hall, it would also serve as a natural divider between the living room and the entry.

Don also readily agreed to extend more base and wall cabinets into the dining area. These would provide storage for extra china and silver and their counters would be ideal for buffet serving.

It was hard to believe things were going so smoothly; at last Don and I seemed to have reached a meeting of the minds. It was the first time since some of our very early confrontations that we could look each other in the eye without rancor. Somehow we seemed to be recapturing our original facility for working together harmoniously and effectively.

I should have sensed that this idyllic interlude would be short-lived. Before long something called a "change order" reared its ugly head. The term itself sounds innocuous enough; obviously it meant changing some kind of an order. The phrase kept popping up in our discussions with Don and before long it dawned on us that change orders were not the routine adjustments we'd assumed they were. Each time we added, took away, enlarged, moved, raised, or lowered anything in the plans as they were submitted to the builder, we created a change order—and an additional expense. In the course of planning and building any house, a lot of adding, moving, raising, and lowering goes on. As we studied the drawings and became more familiar with their finer points and as construction progressed, we found ourselves making many adjustments in the plans. While Don pointed out some of these changes might be costly, I had no idea exactly how costly. Nevertheless, it often seemed wiser to correct a mistake *before* it happened than to have to live with it forever after.

As Don explained much later on, many minor changes were actually made in the plans without any additional cost to us. This had apparently been accomplished through the combined good graces of Don and Dick Riggio. Don had successfully bargained on our behalf for many of these substitutions. In other cases, although the adjustments might have added to our overall costs, Dick Riggio apparently sustained the increases purely out of his generosity and good will.

Still, many changes accumulated as we reviewed and refined the plans—changes which made substantial inroads into our budget. Some of these were due to my own second thoughts or sudden impulses. My inspired idea to widen the decks for more gracious living had netted us a surcharge of $1,029. Our firm resolve to use quarry tile instead of resilient flooring and the decision to extend the tile around the fireplace and hearth came to a whopping $1,149. The 16-foot banquette base (sans upholstery) would cost us another $240. Increasing and extending the kitchen cabinets into the dining room would add up to $210 more.

These change orders also covered many expenses which we'd foolishly assumed were part of the plans submitted to the builder: a wall-to-wall mirror and medicine cabinet in our dressing room, $255; boarding or flooring for Steve's deck, $200; driftwood paneling for Steve's bedroom and sitting room, $558; additional storm doors and windows, $417. All this would add up to almost $3,000! But we had just started to build and this was only the beginning.

From this time on we'd quake at the sight of Dick Riggio's pink change order sheets. But the "changes" kept coming in—

$350 for labor for putting up wallpaper, $110 for painting garage walls, $75 for an outside planter and $163 for capping the old concrete seawall crossing our property to cover up the rusted and broken iron spikes.

A few items relating to solar heating brought more change orders: $390 for metal caps at the top of the collectors; $195 for three removable panels to provide access behind the collectors; and, lastly, a miniscule $15 for boarding up the area under the eaves where Don had originally planned an outside roll-up screen for better protection from the hot summer sun.

The final tab came to a walloping $6,469 over and above the builder's estimate. Another large chunk of this was an "allowance" for a washer, drier, and dishwasher. It was that word that threw me. How could I know that in a building contract "allowance" bears little similarity to the dictionary definition of the word? In builder parlance it indicates that (a) the price is listed separately, (b) the owner may choose the fixtures and appliances within the specified "allowance" figure, and (here comes the clincher) (c) the owner will be billed for the full cost and, hence, "allowed" to pay for it. It wasn't until we'd moved into our house that it suddenly dawned on me that our allowance really hadn't "allowed" us a thing.

Here again, a basic lack of communication between Don and me led to my gross misapprehension and mistaken belief that we would be allowed to select all our appliances up to the amounts specified—at the builder's expense! Somehow, neither of us seemed capable of laying these matters on the table so they could be discussed openly and freely, and the air could be cleared once and for all. Instead, each of us seemed to silently harbor our own resentments and grievances about the way the budget was being handled.

C.B. viewed the whole matter quite differently. He thought I should have realized right along that any changes in the plans would necessarily increase our budget, especially when extra material and labor were involved. He was convinced that Don could not have had any way of estimating most of these expenses beforehand. I conceded that this was indeed true in some instances. For example, since the heating contract was not prepared until mid-November, there was no way that Don could have anticipated the need for caps on the collectors or how much plywood would be needed to close in the storage tank.

While we struggled with these financial problems, the house continued to take shape—sometimes much larger and more "shape" than I'd anticipated. One day, as I was tentatively exploring the unfinished depths of our garage and basement, I ran headlong into an enormous "thing." At first I thought a tractor-trailer must have lost its rear end. But what would a tractor-trailer be doing in our garage? This thing was metal and rounded; it was taller than I and over 12 feet in length. If it had ever been in Don's drawings, I for one had never seen it! (How could I ever forget it?)

At this point the basement was still a gaping hole with only the cement-block side walls and a few studs standing above the foundation. Although the main part of the house had already been closed in, the garage was still without doors or flooring. Now, I began to wonder, how could this colossus ever be accommodated in the nether regions of the house?

Surprisingly enough, it did fit. When we arrived the following weekend, this sleeping giant was nestled into one side of the basement, extending itself into the back part of the garage like some huge space-age monster. But now that it was stashed away, how in God's name would we ever get it out without tearing down the whole house?

I could scarcely wait to call Ev about this basement behemoth. When I did, he patiently explained that this was none other than our solar storage tank. He assured us that it would soon be walled in and totally out of sight. Furthermore, he planned to build a special door to provide access both to the tank and to the other solar controls in the area.

By my next visit, the tank had indeed disappeared. It was totally out of sight. The trouble was that there were some other things totally out of sight as well. Like the two large storage closets which Don had promised us: one for household cleaners and supplies and the other for outdoor equipment. Evidently the giant solar cylinder had taken these closets along with it when it vanished from view.

Ev was apologetic about usurping all this storage space. He said it was the only place the tank could be accommodated within the four walls of the house. As a consolation prize, Ev and Don later did carve out a fairly large closet in the garage for C.B. to store his boating gear.

We did not realize that this was only the beginning of the solar invasion of our inner space, and only the first of several closets to be given up. Nor did we yet appreciate that anything solar always seemed to take precedence over all other requests, despite our most fervent supplications to the contrary. I now know what was meant by sacrificial offerings to the sun.

Our solar storage tank and its thermometer later became great sources of interest for our visitors. On a sunny winter day the thermometer, which registers the average temperature of the collectors, often reads 135°F. In more moderate weather the average collector temperature will be 150°F. or more. Many people find it hard to believe their eyes.

But I was still troubled by the thought that some day we might have to tear down the house to service or repair the solar storage tank. Ev assured me again that he had left enough room for easy access to the tank. If necessary, he told us, he could always remove one of the plywood panels enclosing the tank.

We couldn't help thinking how fortunate we were that Ev was so slight and agile. If Don or C.B. had been our solar engineer, there'd be no way either of them could ever get within arm's length of that tank. We had discovered a new solar principle: be sure to choose a solar engineer who is agile, adroit, and not too tall.

Don had been keeping us apprised of Ev's progress during this early phase of construction. These periodic bulletins had been very encouraging and they reassured us that Ev was hard at work and making headway with our solar heating plan.

We were quite surprised, therefore, when Don announced in early November that Ev wanted to meet with us as soon as possible. Don hinted that Ev had run into some serious problems and so we arranged to meet at Don's office the following weekend.

As the week went by, our apprehension mounted. Had our friends been right? Was Ev going to give up the whole idea? Maybe we should have taken the MIT solar homes more seriously: after all, none of them existed any more. Ev had never submitted the heating plans and we had no formal agreement with him; nor had we paid him a cent.

At the meeting, C.B. and I were both greatly relieved to find that it was not the solar plan itself that was causing Ev's problem. But it was clear that he was most unhappy and ill at ease. It was the first time we had seen Ev downhearted and presumably at a loss for words. After some hesitation, Ev came out with it: he had neither the money nor the help to build our solar heating system! His associates had pulled out and wanted no part of our project.

C.B. took a quick assessment of the situation, asking Ev a few questions. C.B. seemed to grasp the nature and extent of the problem immediately. He first wanted to know how much Ev would need to finish the plans for our solar heating. But Ev explained that it wasn't just a matter of finishing plans: once they were completed, he would need a tremendous amount of raw materials and fittings, and he did not have the capital for these expenses. For example, he would need over 400 square feet of copper sheeting alone, to say nothing of all the coils, pipes, pumps, fans, controls, and other necessary units. When C.B. tried to pin him down to a specific figure, Ev finally said he'd need around $3,000.

Without hesitation, C.B. offered to advance the full amount; he said he would consider this a down payment on our heating system.

Ev seemed stunned for a minute. When he realized the implications of the offer, his face lit up like that of a little boy who has discovered his favorite toy beneath the Christmas tree. Ev's appreciation was probably due as much (if not more) to C.B.'s display of unqualified confidence in him as it was to the financial offer.

Although obviously very relieved and pleased with this development, Ev was still confronted with the loss of both his technical assistants. When C.B. pursued this subject, Ev admitted that, given some time, he could probably enlist a few technical education teachers in the area to help him complete and install our solar components.

What an unexpectedly favorable turn of events! After Ev had briefly reviewed the system he was planning, he gave C.B. the agreement he'd drawn up and we all adjourned for a victory libation.

6

The Solar Heating System

November 28, 1973, was an important date for the C. B. Franks, and perhaps a historic date in solar home-building. On that day we signed the solar heating contract with Everett M. Barber.

All of us were jubilant. C.B. had read the contract carefully and was very pleased with it. The financial arrangements between Ev and C.B. had worked out to their mutual satisfaction. And Ev had been able to convince several local mechanical contractors to help him with our solar project.

The heating contract was a two-page typewritten letter addressed to Mr. and Mrs. Clarence Frank. It was brief and to the point, yet thorough and specific. The uncomplicated and direct language indicated that Ev had drawn it up without benefit of legal counsel.

According to the contract, Ev would install "a solar collector-thermal storage system" to provide the heat for a forced-hot-air heating system. This would include: (a) a conventional 50-gallon oil-fired domestic water heater to be used as a backup for the solar system; (b) thirty-six 2-by-6-foot "solar collector modules" which would be integrated with the roof; (c) a 2000-gallon heat storage tank (with which we were already acquainted); (d) a 275-gallon fuel oil storage tank; and (e) an air-handling unit, including coils, motor, filters, and fan. When installed, the solar heating system would also provide "all supply and return ducting, registers, grills, dampers, as well as the related piping, valves, controls, fittings, and pumps." All pipes would be well insulated and all ducts and fans would be adequately sized for the possible addition of air conditioning later on. Another important clause stated that the system would be planned to provide adequate domestic hot water.

Certain plumbing and electrical matters were excluded from the contract, such as the connection of plumbing lines from the main entry to the water heater and fixtures, and the electric service and connections to the fans, pumps and controls.

According to the contract, our solar heating system would be capable of maintaining 75°F. "in the occupied spaces" when the outdoor temperature was 5°F. The total cost of the system was "not to exceed $7,000."

Ev took the opportunity to explain additional details about the design and operation of our planned solar heating. For the first time we learned that there were other kinds of collectors which Ev could have decided to construct. One type, known as a concentrating or focusing collector, could magnify the sun's heat so tremendously that it could achieve temperatures of 300°-400°F. or higher. There were also tracking collectors that could automatically (or manually) follow the course of the sun to capture the maximum amount of solar radiation at different times of the day and during different seasons of the year. However, Ev explained, these collectors were very costly to assemble and would probably require considerable maintenance. They would also need an especially large area for mounting. We were astonished to learn that even barrels and bags of water had been used successfully as rooftop solar collectors, especially in the Southwest.

Ev told us that in our case concentrating collectors would not be necessary since solar home heating operates well with temperatures of 120°F. or less. Ev had therefore decided to build flat-plate collectors. We seemed to recall Ev telling us earlier that these collectors had the advantage of being able to utilize diffused as well as direct sunlight. This feature would enable our system to collect solar energy even on days when there was no bright sunshine.

Because the south side of our house would be almost entirely glass, we would be able to collect enormous amounts of solar heat passively, without any mechanical assistance. This too, Ev said, would add considerably to the effectiveness of our solar operation.

By now we had a vague idea of what our collectors would look like. Ev had worked out their design and knew exactly how he planned to build them. The thirty-six collectors, consisting of oblong boxes only 6 inches deep, would be mounted side by side in three parallel rows across the roof. There would be no need for special bases since the roof design allowed for the collectors to be mounted between the rafters. Once mounted, they would be sealed in and would become an integral part of the roof itself. The mounting, we learned, had to be done carefully. An exact fit would be of critical importance in collecting and retaining solar heat.

Ev went on to explain that the greatest amount of solar radiation can be collected when a surface is at right angles to the sun. Therefore, at the equator, where the sun is directly overhead, the best position for collecting solar energy would be horizontal. But as you travel north into higher latitudes, the sun is lower in winter and the collectors have to be positioned increasingly more upright. The rule of thumb for calculating the best collector angle for home heating is to add 15° to the latitude of the particular location. As Ev had told us earlier, the optimal angle for our collectors would be a 57° tilt. Since flat-plate collectors are stationary and do not follow the sun, the angle selected would necessarily be something of a compromise. Our collectors would be specifically slanted or tilted for collecting the greatest amount of solar radiation during the cold winter months.

For the first time, these solar concepts and principles were beginning to make sense to us. We could now understand that if we lived much farther north—in Alaska, for instance—our collectors would have to be almost vertical in order to be at right angles to the sun and thus collect the greatest amount of solar energy during the winter heating season. On the other hand, if we lived much farther south and much nearer the equator—in Arizona, let's say—our collectors would have to be tilted much closer to a horizontal position.

We were also beginning to appreciate our beach location where we wouldn't need to worry about trees or tall buildings. Now we could understand why high buildings, especially in congested areas, could seriously interfere with the collection of solar energy on nearby, lower buildings. The extent of this interference would vary depending on the latitude, the time of day, and the season of the year, all of which determine the sun's position in the sky.

Up to this time we had not been aware of the big boost our solar system would be getting as a result of our unobstructed shoreline location. With so many miles of Long Island Sound's waters south of us, we would inevitably gain a great deal of reflected radiation from the strong sunlight shining on the surface of the water.

Ev had already begun to fabricate some of our thirty-six collectors, each to be constructed entirely by hand. He had chosen to fashion the flat plates of copper because it is easy to work with and is an excellent conductor of heat. While aluminum would have the advantage of being lighter and cheaper, it would have been more susceptible to corrosion. Galvanized iron or steel would also have been cheaper, but would not have conducted heat nearly as well as copper. Nor would either of these metals have been as durable as copper.

The collectors would need to be completely dustproof, moistureproof, and airtight to reduce possible heat loss; each of them would have to be carefully prepared to fit perfectly. In choosing the material for the collector covers, Ev would also have some options. While plastic is cheaper and lighter than glass, it has a tendency to become brittle and yellow with age. If it is exposed to high temperatures for a long time, it is likely to soften.

Ev had, therefore, decided on glass covers. While the risk of breakage would be far greater with glass, it doesn't scratch as easily as plastic, and it would probably be more durable in our exposed location. Ev would need enormous amounts of glass to provide covers for every one of the thirty-six panels, each with an area of 12 square feet.

Ev had also prepared a special black selective coating to be applied to the copper backplates. He was convinced this coating would promote greater absorption of solar radiation adding to the efficiency of our system.

Between the glass cover and the absorbing plate of each collector would be a layer of air to provide additional insulation and reduce the loss of solar heat once it was absorbed by the copper sheet. A 3-inch backing of fiberglass would contribute further insulation.

Ev still planned to use the propylene glycol solution, which was nontoxic, as the medium for transferring solar heat from the collectors down into the storage tank.

Until this time, we'd had no idea that collectors didn't always operate with fluids. Ev explained it was also possible to use air as the medium for transferring heat. He went on to say that air has several distinct advantages. First, it is far less corrosive than almost any kind of liquid. We'd also never have to worry about having the collectors freeze. Finally, the construction of the collectors would not need to be so exacting and painstaking. A leak from an air collector would be far less damaging than a leak from a liquid system. Ev said that rocks or stones are also often used to store solar heat in air systems, usually in a large bin in the basement area. In our case, Ev believed liquid collectors would be most practical and efficient.

To calculate the exact number of collectors we needed, Ev had taken into consideration many of the particular climatic conditions in our area: the temperatures we could anticipate, based on records of past winters, the amount of sunshine we could probably count on, the direction of the prevailing winds, and the expected wind velocity. This information, Ev told us, was available from the National Weather Service; many local and state agencies could also be helpful in these calculations. Such things as altitude or the presence of fog and other pollutants in the atmosphere or other local characteristics could interfere with or enhance the collection of solar energy. For example, at the seaside, while we would have to contend with frequent fogs, we would be less likely to have any serious air pollution.

Ev estimated that the total area of our collectors would be 430 square feet. He based this figure on the various climatic factors, expected efficiency of our collectors, and the relationship between the amount of space to be heated and the size of the collectors. The surface area of our collectors would be less than a quarter of the 2,000 square feet of interior space to be heated, roughly a 1-to-4 ratio. The early MIT homes had required a much higher ratio of 1 to 2, but improvements in collector efficiency have made it possible to get adequate solar heat from smaller collector surfaces. In our case, limiting the surface of the collectors to 430 square feet was also a matter of necessity. Since our house was multilevel, we had a relatively small roof and the collectors would cover almost all of it.

Ev was confident we would be able to derive at least 45 percent of our heat from the solar system. (Actually, we have gotten closer to 65 percent!)

Since our collectors would be mounted at a 57° angle, the chances of dirt accumulating on them was relatively slight. Any material which might cling to the glass covers would probably be washed down by rain or snow. If any snow happened to adhere to the glass, the heat of the circulating glycol solution would quickly melt it. Ev suggested we might find it advantageous to wash the glass panels once a year.

Our collectors, Ev said, were being especially designed to withstand high winds and bear the weight of heavy snow. But no matter how many precautions we might take, it would never be possible to provide absolute protection for the collectors under all kinds of weather conditions—no more than we could provide absolute protection for the house itself. Damage from hurricanes, hailstorms, or other "acts of God" were natural hazards to which the collectors and the house would both be vulnerable. (Of course this would be equally true for any other house in our shorefront location, solar or otherwise.)

By far the best and only way to safeguard the house and the solar equipment from major catastrophe would be to provide full insurance coverage. Up until now we had never thought about the need for insuring our solar equipment. Ev had spelled out some of his guarantees in the contract, but even he had to exclude "acts of God." It would be interesting to find out where insurance companies would stand on new solar issues. It wasn't likely that they'd ever had any previous experience insuring solar heating equipment.

(Later, when we applied for home insurance, there was some discussion about the construction of the collectors and the danger of breakage. Our insurance agent decided, however, that the collectors were an integral part of the house and could therefore be included in and protected by our regular homeowners' policy, with no increase in our rates. We had already planned to have maximum coverage for wind damage. Coverage for water damage was available to us under a special federal provision for residents of low-lying, flood-prone areas such as ours. Our solar equipment is now as fully insured as any other part of the house.)

We faced one other hazard peculiar to our coastal location. Seagulls have a rather messy habit of dropping clams and oysters from great heights to crack open the shells and get to the succulent and delectable innards. This approach, while often more successful than our own clumsy methods of opening shellfish, did pose a potential threat to our glass-covered collectors. While the 57° slant was likely to deflect most objects, the gulls' clams-on-the-half-shell could become a real problem. If this happened, Ev advised us to cover the glass panels with chicken wire. But because this screening could cut down on the efficiency of the collectors, he recommended it only as a last resort.

Since our collectors would be made of durable materials without any moving parts, Ev didn't anticipate any major expenses for service or maintenance. Most sections of the solar heating system could be checked and repaired by any competent plumber, provided he was given a briefing on its structure and operation. Ev had already volunteered to service our heating units without cost for at least the first year. This proved to be a more generous offer than we thought at the time. Although we have never experienced any major difficulties, there was literally no time of the day or night when Ev did not respond promptly to our calls about the solar equipment. When necessary, he would arrive in person within minutes. Ev never charged us for any of those endless hours he spent checking and rechecking our system, making minor adjustments, monitoring the performance, and adding or taking away fittings to improve our solar operation.

As both Don and Ev emphasized, many energy-saving features had already been incorporated into our house: the compact design, the recessed entries, the double-glazed windows, the limited number of openings on the north side, the extra insulation—each would contribute to the efficient functioning of our solar system. The fiberglass and the thick layer of polyurethane sheeting over the plywood on the outside walls, as well as the air spaces left between the different layers of insulation would also be effective in decreasing the penetration of cold and moisture. Our white cedar siding would be the frosting on the cake, providing an excellent outside barrier against the cold. And our open interior plan would promote the circulation of air and heat so they could rise freely and easily into the upper chambers.

Also, during construction Dick Riggio would be taking great pains to see that all doors and windows fit snugly and were well caulked and weatherstripped. Later, we decided to add storm doors and storm windows wherever the builder had not provided them. These would serve as additional barriers to heat loss and to the infiltration of dampness and cold.

Since heat is easily lost through the roof, this area would be given special attention and extra insulation. And, Ev explained, all the pipes and ducts connected with the solar heating would be swathed in heavy batting to cut down losses while the heat was in transit. Even our giant solar storage tank, peacefully slumbering in the basement, would get a heavy dusting of vermiculite to provide further insulation.

The combination of these energy-saving features could reduce our overall heating requirements appreciably, thus decreasing the amount of heat the solar system would need to produce. The combined cost of all insulating materials was not insignificant and would add to our initial capital investment. But Don and Ev convinced us that these measures would, in the long run, pay for themselves in fuel savings by reducing our need for oil heat.

Our heavy insulation had a surprise bonus. While the primary function of insulation is to keep houses warm during cold weather, it is equally effective in keeping houses cool and comfortable during hot weather. As a result, we later found no need to install air conditioning.

When it was time to select more furnishings, Don suggested we use especially heavy, lined draperies, shades, or shutters inside all glass areas to help retain heat, particularly at night. After considerable exploration of various possibilities, we decided on Roman shades, some wood and some fabric, both of which have good insulating properties. When rolled up horizontally, these attractive devices remain completely above all glass areas, depriving us of not even a smidgeon of our precious views; they have the added advantage of being very inconspicuous when not in use.

Our solar system was designed to supply more than half of our domestic hot water—another effective way of decreasing the need for our backup oil heat. With our automatic dishwasher and washing machine, Ev estimated that close to one-third of our total heating supply could be needed for our domestic hot water alone.

Solar hot water heating has one distinct advantage over solar space heating. In the typical household, the need for hot water is usually greatest during the summer months and coincides with the time when the greatest amount of solar radiation is available, whereas the need for space heating is definitely greatest during the cold winter months when the least amount of solar radiation can be collected. Solar heating of our domestic hot water would, therefore, be an excellent way of utilizing our collectors on a year-round basis.

We began to hear more about the prevalence of solar water heating abroad and the use of solar panels to heat swimming pools in the United States. This practice was growing rapidly, especially in states like Florida, where swimming pools are often looked upon as essential adjuncts to a house. (At this writing, Florida and several other states prohibit the use of conventional fuels for heating swimming pools. This will unquestionably lead to a rapid rise in this application of solar heating.)

Our solar hot water heating system would be more sophisticated than some of the rather primitive units used abroad. It would also be more complex than most of the simple passive solar heating units being used in other parts of the country. However, the underlying principles would be the same.

Unfortunately, later we ran into some unexpected problems with our hot water heating. At first, the local sanitarians were concerned about the rather remote possibility of infiltration of the propylene glycol solution into our drinking water. After that was worked out, Ev found it necessary to add an anticorrosive which *was* toxic to the solar storage tank and this ultimately led to some major changes in our hot water system.

But all that was far in the future. For the moment, we could rejoice in the knowledge that our solar contract was at last signed, sealed, and delivered!

7

Coping with a Solar Dragon

The word was out: the first of our solar collectors would be mounted on our roof on January 17, 1974. Much to our surprise, that event caused quite a bit of fanfare. We'd no idea how rapidly the interest in solar energy had been growing over the past year. We were amazed on that bitter cold morning to find reporters, photographers, and television crews appearing on the scene to cover the big story.

Photo by Robert Perron

While Ev and Dick Riggio perched precariously on the roof, Don handled the growing group of spectators. After a few tense moments, the first collector was raised into the air and then slowly and gently lowered into place. As the collector settled down into its new and permanent home, the audience shouted enthusiastically and burst into loud applause. It was a memorable moment.

The atmosphere was one of joyous celebration. Ev's wife and loyal supporter, Ginger, served hot coffee and doughnuts. Several neighbors arrived trying to find out what was going on. This was the day we had been so eagerly awaiting for such a long time.

Although the house plans and the model had received much attention from the local press, this was the first time we had encountered reporters from city newspapers. The media people were eager to find out all about us and our plans to build a solar house: How and when did we first hear about solar heating? When and why did we decide to go ahead with it? How long would it take to build the house? How much would it cost?

Until now, we hadn't considered the house of any importance to anyone but ourselves. Despite the worsening energy situation, most people we knew still considered our house an interesting experiment at best.

Now, for the first time, we realized our solar house was causing more than a passing ripple—the skeptics and cynics notwithstanding. The predominantly young people from the media seemed to share a strong belief in the future of solar energy. They were genuinely interested and excited, hailing us as solar pioneers and equating our solar house with earlier inventions of historical significance. None of them seemed to have the slightest doubt that solar energy was the wave of the future and that our house was only the first of many thousands of solar homes sure to follow.

I suddenly found myself somewhat embarrassed by their sincere and repeated congratulations, especially when I recalled my initial resistance to the whole idea. It was C.B. and Don and Ev they should be congratulating, not me.

By March 1974 our solar house was closed in. Construction seemed to speed up in the weeks that followed. And along with this, our problems as home-builders took on new dimensions.

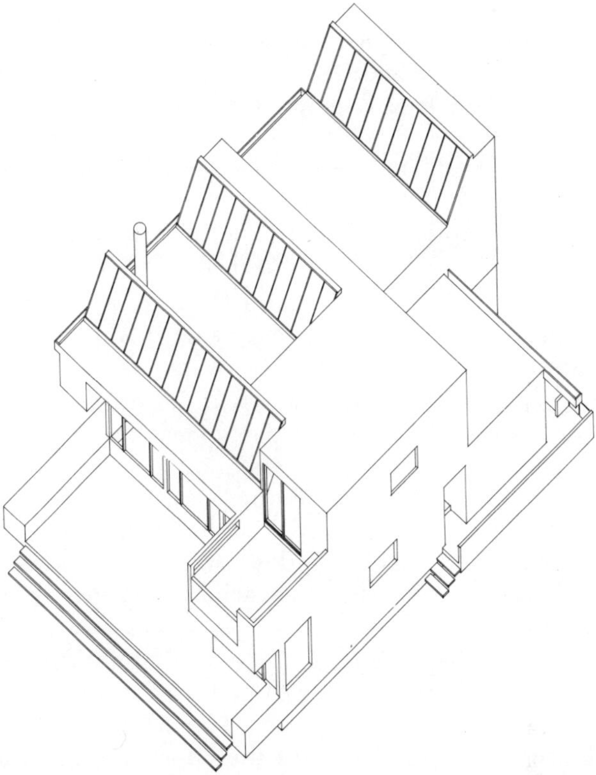

Courtesy of Donald Watson, AIA; from *Designing and Building a Solar House*, Gardenway, 1977

Sol-A-Sea's solar collectors are mounted in three south-facing rows

Before this, the problems had been on paper. Don and I debated over erasing this and sketching in that. Now when I ran into problems, they were real problems, like the day in early April when I ran into a wall. A real wall, a wall that had never been there before, a wall that had been . . . my broom closet!

Closets and storage space were of great importance to us at the time. We would be consolidating our large ten-room house in Stamford into our much smaller beach house. There wasn't too much difference in the overall square footage, but the beach house had much larger rooms and, therefore, much smaller storage areas. We would need storage space for the many things we'd accumulated in over thirty years of marriage.

The more I thought about this, the angrier I was. Then I became suspicious. Had any more of my closets disappeared? I made a complete tour of the three living levels, ending in Steve's quarters. Sure enough, something was missing: one whole corner of the bath had been boarded up. But what had been there? Steve's linen closet! It was gone, swallowed up by some solar dragon. (I soon found myself becoming obsessed with solar puns. My favorite song was "O Solar Mio." The closets were being eaten by solar dragons. A high change order became a low blow to my solar plexus. An unpleasant surprise became a solar shock.)

I decided to check the plans and make sure the closets had existed in the first place. When I found this to be the case, I still hesitated to call Don. After all, what could he do about it? Make them rip up their new construction?

Later, I called Don anyway, just to let off some steam. He was delighted to be the innocent party for a change. He quickly passed me on to Ev, who was understanding and sympathetic, as always, but immovable. He explained that he needed the closet space for the vertical risers which would bring heat to Steve's area. With our tight planning, there was just no other place to put these big ducts. Some were as much as 30 inches wide. I'd lost another round!

When we met with Don again, he tried to appease me about the closets. He was sure we still had plenty of storage space. C.B. didn't offer much comfort, either. He said he didn't consider Don's robbing us of two closets for the solar equipment a capital offense. Again, there wasn't a thing I could do.

The case of the disappearing closets was followed only two weeks later by the case of the dismembered basement. From the beginning we'd planned to divide the basement into a good-sized workshop for C.B.; a large area for family storage; an "engine room" for the heating apparatus, fuse box, and so forth; and, lastly, the main and largest area, a space to be reserved and finished later on as a dormitory and recreation room for our children and grandchildren.

When I peeked into the basement in early May, only one of these areas was still exactly as planned: the space set aside for the recreation room, later known as "Shipwreck Alley." C.B.'s workshop and the storage area had been combined and were half the size of what I'd wanted. All other space had been gobbled up by the engine room. It was just too much! If this kept up we'd probably end up storing our belongings on the seawall. When Ev arrived later in the day, he offered his sincere apologies for carving up our basement. But, he said, he just had to have additional space in the engine room for all the solar equipment.

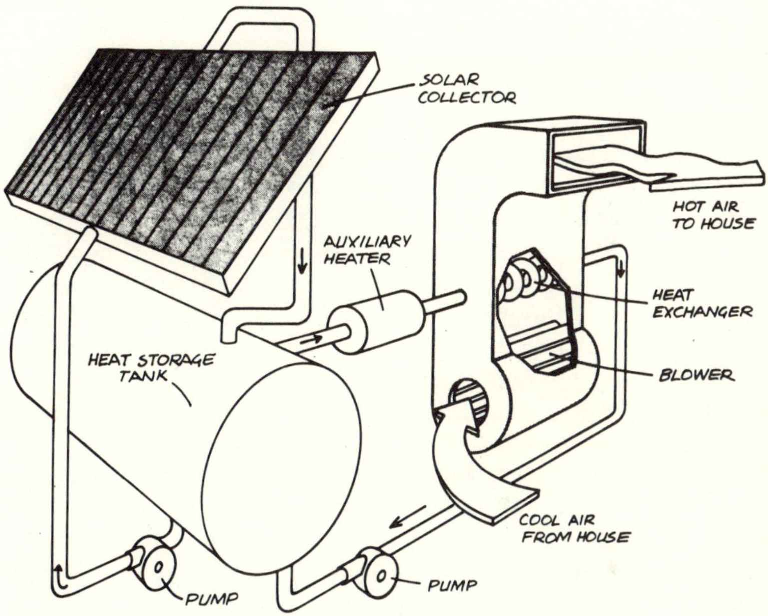

Typical active solar-heating system

From *The Solar Home Book* by Bruce Anderson with Michael Riordan, Brick House Publishing Co., Inc., Andover, Mass., 1976

While I wasn't very sympathetic at the time, later when I saw what the engine room contained, I had to concede that the space was needed. Between the hot water heater, the oil storage tank, the stacks of glass panels and big black filters, and the enormous metal containers (holding blowers, circulators and other mysterious apparatus), every nook and cranny was taken. The walls and ceiling were covered with an unbelievable maze of vertical and horizontal pipes, crisscrossing each other in a totally incomprehensible fashion—pipes running up to and down from each row of collectors as well as to and from the water storage and oil tanks. In addition to all this, the room contained the usual household plumbing for a 2,000-square-foot house.

One of the most dramatic encounters over changes to the house was not with Don or Ev but with a four-year-old boy. One day, when I happened to wander into Steve's room, I found Ev's adorable, wide-eyed son busily sawing a hole in the floor. I must have let out a blood-curdling shriek. The poor child burst into tears and ran downstairs bellowing, "Daddy!" Daddy got there in a hurry and explained that this little fellow had only been carrying out his orders. Ev had carefully marked the spot for one of our heating outlets and his son was only helping him prepare the floor for it.

Although I felt very much like the wicked witch of the West, I was glad the hole in the floor would serve a useful purpose. Later, Ev's son and I got on a better footing and I think he has forgiven me that initial indiscretion.

Despite these minor catastrophes, the house was shaping up and we were beginning to realize what a light, airy, and beautiful place it was going to be. Seeing each room emerge as an entity with individual characteristics and charm was rewarding. Even without carpeting, the living room had become a cheerful and inviting area. The cedar paneling, stained in a soft, warm driftwood finish, was a perfect complement to the large tile areas surrounding the fireplace. We were waiting impatiently for the base of the banquette to be finished so we could order upholstery material, and we were looking for some simple swivel chairs and a large round cocktail table to pull the room together.

By now the cabinets for the kitchen and dining area were all in place and the tile flooring had been laid. We needed only our major appliances for the kitchen to become operational.

The carpenters had recently been spending their time on the built-in furniture for the master bedroom. We were delighted with both the material and the quality of workmanship. Don had persuaded us to dispense with hardware on all the cabinets and drawers, which gave the built-ins a sleek, contemporary look. (They were easy enough to open by pulling on invisible indentations carefully routed out beneath each door and drawer.)

One afternoon I noticed a crack across the top of C.B.'s handsome chest of drawers. No, it wasn't a crack; it was a hole, a gaping hole! Someone had performed major surgery on C.B.'s beautiful, brand-new, custom-made chest of drawers.

As I stood there dumbstruck, I began to wonder—could there be holes and fresh scars on the other pieces of furniture? A quick survey confirmed my worst suspicions: the desk and triple dresser had even larger and uglier holes, 18 to 20 inches long and 3 to 4 inches wide. This was no carpenter's mistake; someone had deliberately hacked away at our fine teak treasures. This was the bottom, the last straw. It was intolerable and I refused to accept it. I raced upstairs, where Ev was working, and demanded an immediate explanation.

As always, Ev was all apologies, but immovable. He patiently explained that the holes were heating registers, which had to be where they were. There was just no other place to put them. Our built-ins abutted the walls and the heat came through the walls. The registers, therefore, couldn't be near or behind the furniture or even underneath it. The only alternative was to go *through* the furniture.

I could feel myself flushing as the rage built up; tears were coming fast and would soon spill over.

Ev was upset. He was sincerely sorry, but what good did that do? He kept muttering something about how he'd cover up the holes with grilles, how we'd hardly notice them. But we already had noticed them and we would continue to notice them.

Within a few weeks these unexpected and unwanted openings were screened with some less than attractive grillwork. By then, they had become catchalls for paper clips, pens, bobby pins and the sort of debris that accumulates on the tops of desks and bureaus. Eventually we did become reconciled to our "hole-y" furniture and did not notice the holes . . . too often.

Other heating outlets kept popping up in the most unlikely places: one directly in front of the sliding glass doors leading to Steve's deck and others under his captain's beds and campaign desk (also custom-made for a perfect fit). Eventually, we were able to make adjustments by placing the furniture a few inches away from the walls so the heat could get through the registers.

Fortunately, the heating outlets in the rest of the house did not present serious problems. Several large wall outlets took us a while to get used to. But before long we had them painted to match the surrounding walls and they have become quite inconspicuous.

Other construction was progressing nicely. Don had designed a 4-foot deck overhang to protect the south side of the house from the direct heat and glare of the summer sun. This also provided some protection from heavy winds and rainstorms. He had made careful calculations to be sure the overhang would not interfere with the infiltration of the welcome winter sun. Since the sun is lower in the sky and traveling a more southerly route in winter, it could still penetrate the southeast, south, and southwest glass sections of the house during the heating season.

Don also agreed, somewhat reluctantly, to connect the main deck on the beach side with the narrow side deck and walkways going around the house. This was important because it made it possible to walk completely around the house without having to go through it. While expanding and connecting the decks did result in one of our costlier change orders, it was one expense we never regretted.

Other details were finally falling into place. I had to agree that the firebox with tile treatment on the surrounding areas made the fireplace charming and attractive. But I was still not completely reconciled to the 5½-foot exposed shiny metal chimney and felt it introduced a jarring and discordant note into the room. Then one day the painter generously offered to paint the galvanized chimney to match the walls. Once painted, the chimney became an attractive asset to the room and now I find it quite charming. It does add another interesting architectural form and has created a high, wide setback above and around it.

With the design of the decks and the living room settled, we looked forward to sailing along to a smooth finish.

As it turned out, there were still a few surprises in store for us. One of these occurred when Bob, Riggio's talented head carpenter, proudly displayed the handsome base he had just completed for the 16-foot banquette. It was beautifully crafted of stained cedar paneling, and Bob was especially proud of the large storage sections beneath the seating. It was an elegant and useful foundation, but it was in the wrong place! Instead of being 18 inches in front of the wall, the banquette base was built directly into the wall. This made it impossible for us to have the floor-to-ceiling bookcases we'd always planned for that wall. Unfortunately, since this was one of the details omitted from the base bid, the carpenters had had no way of knowing that the bookcase was supposed to extend eighteen inches into the room behind the sofa.

We tried to conceal our disappointment and distress from Bob, but it was difficult to be enthusiastic. There wasn't much point in tearing down the base, so we resolved to live with it. Later, it proved to be a blessing in disguise. If the banquette had been further forward, as planned, it would have made the living room look narrow and cramped, with the sofa dominating the whole room.

By now, our house had developed a personality of its own and it was high time to choose a name for it. But what would be appropriate? The name would need to reflect the solar features of the house and also give some indication of its beach setting. And so we christened it "Sol-A-Sea." The name (and the house) has achieved far more publicity and stature than we have—which is really the way we hoped things would work out.

But before we could call Sol-A-Sea home, we had a few more solar surprises. Late one day, I ran into Ev as he was rushing out of the engine room. "Oh, by the way," he said, casually, "have the painter leave enough extra paint for the living room ceiling, in case I have to break through it."

I made a supreme effort to keep cool. Could he be serious? Would he joke about something like this? I demanded an explanation for this latest threat to my equilibrium.

Ev tried to explain that this would be the only way to gain access to the second row of collectors, in case they needed adjustment or repair. And then he added, "And I might have to go through the ceiling in the master bedroom too. That'll be the only way I can get to the third row of collectors."

This was the ultimate insult, the final indignity. I thought I'd become inured to these solar shocks, but I was so angry I couldn't say a word.

As Ev and I stood there glaring at each other and heading straight for a major confrontation, Dick Riggio appeared on the scene. Ev explained the problem and Dick understood perfectly. Within seconds, he took over the situation. He was sure he could work out some compromise satisfactory to both of us—an arrangement that would give Ev the access he needed without the necessity of tearing out the ceilings.

Dick soon followed through on his mediation. The very next weekend he stopped by to offer us a solution. He would construct a series of wood panels inside the house, behind and beneath the collectors. He would then stain them to match the woodwork. In this way Ev could easily get to the second and third rows of collectors without disturbing anything. Dick did such an excellent job on the panels that unless we point them out to visitors, no one seems aware of them.

There were other adjustments to be made. When the large heating ducts were mounted on the ceiling of the garage and then swaddled in heavy insulation, almost a foot of headroom was lost. Steve and C.B., both six-footers, had to learn to duck at the proper moment and we have all learned never to come tearing into the garage with the luggage rack loaded.

The final battle of my aesthetic preferences vs. the solar necessities was caused by a thermometer, a commercial thermometer which we suddenly found planted smack in the center of the wall I'd planned to use for displaying our favorite works of art. It was a perfect spot where we could enjoy looking at them every day. I was *never* going to enjoy looking at that thermometer. It had to go! Ev appeared just as I was about to tell one of the workmen to remove the thermometer.

Ev realized the problem immediately and seemed embarrassed and uncomfortable. "Gee," he sputtered, "there is just no way I can disconnect this." He walked over to the wall and with a flip of the wrist, he disengaged the thermometer.

Sure enough, the frame hid a hole with copper tubing in it. The tubing, Ev explained, was connected to the collectors up on the roof and would provide a constant reading of their current temperature. Another smashing solar defeat.

Although Paul Riggio, Dick's son, and Ev offered alternative suggestions, none seemed quite suitable and we decided to live with it as it was. While the thermometer does not have the same appeal as some of our prized paintings and wall hangings, it has become a solar conversation piece. Our guests are intrigued by the high temperatures in our collectors.

If it wasn't thermometers and hole-y furniture, it was missing closets and dismembered basements. But somehow, despite all the problems and delays, by the end of the summer, our solar house was complete. And we loved it.

8

Sol-A-Sea's Fame and Fortune

Exactly one year from the day Richard Riggio of Essex had secured building permit number 5699 to build a solar house at Pilots Point, Richard Riggio of Essex presented us with a certificate of occupancy and we moved into Sol-A-Sea. It was September 19, 1974.

Photo by Robert Perron

While our new house was unusual, our moving experience was typical. Friends and familiarity made us reluctant to leave our old house. Novelty and excitement drew us toward our new house. And confusion and chaos were everywhere.

A few weeks earlier, I had decided to combine a housewarming with our annual "surprise" birthday party for C.B. As one of our friends puts it so aptly, the biggest surprise would be if we *didn't* give the party. Everyone always cooperated and somehow it was always a "surprise" to C.B. But this time I began to wonder whatever had possessed me to try a dinner party for forty people four days after moving into a new house. Unopened cartons were everywhere and there were still a few unfinished spots here and there.

Things seemed to be going fairly well except for the so-called silent toilet, which roared continuously like Niagara Falls. After breakfast on the day of the party, I went to use our brand-new wall ovens but the doors wouldn't open. This might not have been such a tragedy if I hadn't needed the ovens to heat the casseroles I'd already prepared (and hidden) for the dinner party. After stealthily returning the casseroles to the freezer, I called C.B. for help. He tried his hand at the oven doors but gave up quickly. As he took off for the day, his parting shot was, "Oh, well, Ruthie, you don't like to cook anyway."

Later two close friends and neighbors dropped by. But even the combined talents of this eminent pair (a surgeon and a banker) failed to unlock the doors. As a last resort, I called the manufacturer's service representative, who kindly agreed to instruct me over the phone. So with my trusty little oven manual clutched firmly in one hand, and the other hand free to carry out his directions, I was finally able to locate the release buttons. The day was saved. No matter what else went wrong, at least we could eat.

The party was, as always, a complete surprise to C.B. Judging from the "ohs" and "ahs" of the guests, which continued all evening, they all agreed that the house was not only beautiful and unusual but also well planned and very livable. This buffet party for forty people proved how flexible the house was and how well suited for entertaining even such a large group. The guests flowed freely from room to room and the house never seemed overcrowded.

By now our family and friends had become more accustomed to our solar house. Some of them had even become interested in solar heating for themselves. (Later some did build solar homes.) Others were surprised that a solar house could be beautiful and comfortable as well as energy-saving.

Before long, we realized how much interest had been developing in Sol-A-Sea. What started as a small trickle soon became a steady stream of visitors from all over Connecticut, from many other states, and eventually from each of the five continents. Interest in Sol-A-Sea increased in direct proportion to the growing shortage of gas and oil and the steady rise in fuel prices. Almost from the moment ground was broken, people with vastly different backgrounds and interests found their way to our solar site.

Courtesy of Stuart Finley, Inc.

Most of the early visitors were professionals—architects, engineers and scientists—primarily interested in the technical aspects of our solar system. As construction progressed, a growing number of solar enthusiasts and curiosity-seekers from almost every walk of life began to appear. Many were quite serious about solar energy. These potential solar consumers included a rather large number of older people who, like ourselves, were looking forward to retirement. They viewed solar heating as a way to deal with both inflation and the limitations of fixed incomes. Many were far less concerned with the amount of the initial investment than with the projected savings in their monthly fuel bills.

Another large group of our visitors were young do-it-yourselfers who planned to build their own homes and/or their own solar heating systems. They were particularly concerned with the initial cost of solar components and installation and were sincerely interested in finding out more about the construction and operation of our solar system and whether we were entirely satisfied with it. Most of them were already dedicated believers in the future of solar energy and part of a growing sector of the under-forty population determined to protect and improve the quality of their natural surroundings. Each of them seemed to be searching for new methods and approaches that might help them become more self-reliant and less dependent on the products and technologies of our giant corporations. They were very vocal in their opposition to the expansion of nuclear power and very knowledgeable about solar concepts and developments. Their keen interest and spontaneous enthusiasm were delightfully refreshing and reassuring.

One of our early visitors did go slightly overboard in reacting to the house. Ev had asked permission for a group of students from a nearby technical school to go through the house and we had willingly granted it. A few days later, in the middle of the morning, I got a frantic call at work from one of our soon-to-be neighbors, telling me that there were some twenty-five boys climbing all over our house—and that one of them had just fallen off the roof! Luckily, the sand is soft and he suffered no injuries.

While the house was still under construction, even unexpected visitors were seldom a problem. They seemed satisfied to observe the building activity and ask a few questions of whoever happened to be at the site. Even at this stage we found many of our visitors stimulating and enjoyable. Most of them were much better informed than we were about solar theories and techniques. Our roster of visitors soon included people from Europe, Israel, Spain, South America, and Australia. Hearing their solar experiences reinforced our increasing confidence in solar heating.

As long as the house was not closed in, our visitors could come and go as they pleased and it didn't matter if they arrived unannounced. Then we felt no obligation to them; if they arrived at an inconvenient moment, we could excuse ourselves and ignore them without feeling rude or irresponsible.

On the basis of these early experiences, we really could not have predicted what it would be like later. Uninvited guests began dropping in at almost any hour of the day or night, interrupting our routines and activities and invading our privacy. And of course once we moved into the house, we found ourselves in a new and different role. Earlier the architect, engineer, or builder took responsibility for what happened at the site. Now the house was our responsibility. We could no longer be considered merely clients or investors. We had become the owners, occupants, and gracious (or ungracious) host and hostess of the only solar house within a radius of several hundred miles. We were now expected to observe certain social rules and amenities —like greeting even uninvited guests, welcoming them cordially to our home, showing them through the house, pointing out the solar apparatus, and explaining how things worked.

Photo by Robert Perron

The trouble was that Sol-A-Sea was no longer just a solar house. It was now our home. We had made some sacrifices and had incurred considerable expense so we could retire here and enjoy the quiet, peaceful beauty of our natural surroundings. We hadn't built a solar home to become famous and to stimulate interest in solar energy so the whole world would beat a path to our door.

Photo by Robert Perron

Both Don and Ev made a supreme effort to protect us from being swamped by a burdensome number of visitors. In talking with media people they always made a special effort to withhold our name and exact location. This in itself was tremendously helpful in cutting down the number of curiosity-seekers. By screening all other inquiries before giving anyone our address or telephone number, Don and Ev were able to weed out anyone who didn't have a valid and serious interest in solar energy. And they always tried to give us some advance notice before sending anyone to the house. No doubt dozens of visitors were turned away before they ever reached us. However, during our first two or three years at Sol-A-Sea scarcely a week went by without at least one or two visitors, or a telephone inquiry.

Much of the interest in Sol-A-Sea was generated by the extensive publicity we received. What began as a few comments in the press in November and December of 1973 soon grew into national coverage. Before we realized it, Sol-A-Sea had become something of a prototype for solar homes.

Interest in the house began building almost from the first day of construction. Initially, the newspapers from nearby towns carried brief articles about Sol-A-Sea being a "first" in the area. Within a few months the newspapers in larger cities all over the state were carrying news stories, feature articles, and progress reports on Sol-A-Sea. As the publicity spread and more people learned about the house, and as general interest in solar heating grew, we found many people going to great lengths to find us.

One early visitor, with his young family stowed away in the back of his jeep, had driven all the way from Michigan just to see our house. He found us by asking directions to "the solar house" at a local gas station. He arrived late one afternoon and practically demanded to be shown around the house immediately. He was planning to build a solar house of his own and insisted on a detailed explanation of our solar equipment and a full review of our experience building Sol-A-Sea.

One of our most interesting visitors arrived unannounced on a sunny Sunday morning in January of 1974. It was shortly after the well-publicized installation of our first collector and we were entertaining some friends on the rough planking soon to become our deck.

A well-dressed, unfamiliar man appeared as we were enjoying a light brunch—our first attempt to entertain at Sol-A-Sea. Introducing himself as the editor of *House Beautiful*'s Building Manuals, he looked around a bit and then asked if his wife, who was waiting in the car, could join us. They were amazed that most of us had already shed our heavy outer clothing and were perfectly comfortable in the warm January sunshine. Although the temperature was near freezing, the mass of the house acted as a barrier to the north wind, completely protecting us from its chill blasts. Later one of our neighbors, long familiar with our beach climate, arrived in shirt sleeves and shorts.

As everyone was leaving, the editor approached us and said that he was planning a series of articles on solar homes for the fall edition of the *House Beautiful Building Manual* and would like to feature Sol-A-Sea. We had no idea how such a prestigious national publication had heard about Sol-A-Sea but were glad to cooperate. It was good to know we might be able to make some small contribution to the rapidly expanding interest in solar heating. We had always made a special effort to cooperate with anyone who seemed serious about solar heating. Over the years we have found most people thoughtful and considerate of both our feelings and our property. But this wasn't always the case. There have been times when we felt put upon and some people have taken advantage of our situation. Therefore we eventually found it necessary to adopt a tougher policy to protect ourselves from unwarranted intrusions and disruptions. It took us a long time to learn that sometimes the best and only answer is "No!"

The *House Beautiful* article about Sol-A-Sea wasn't published until almost six months after we first met its editor. So much had been happening during construction and there had been such wide publicity that we had practically forgotten all about this project. We were reminded of it when a friend called to say she'd been reading about Sol-A-Sea and wanted more information about solar heating.

The *House Beautiful* article appeared in a special section on energy-efficient homes in the *1974 Fall Building Manual*. It included some interesting shots of the unfinished exterior and a picture of Ev installing the collectors; there was also an architectural drawing of the finished house and a diagram of the solar heating plan. The information was accurate and was presented in a lively, articulate style.

A follow-up article that appeared in the 1975 Spring-Summer edition of *House Beautiful's Building Manual* covered four full pages. In addition to the diagrams shown in the earlier write-up, it included outstanding photographs of the exterior and nine equally excellent interior views showing almost every room in some detail. There was one full view of the front of the house which has since become the standard portrait of Sol-A-Sea.

This photograph—as well as some other shots—eventually became a source of some dismay to C.B. and me. Neither of us knew very much about photographers, their rights and copyrights, or what was considered "cricket" in house photography. We were surprised and somewhat vexed, therefore, when this particular photo began appearing in all kinds of publications—magazine articles, commercial advertisements, calendars, and minor publications by solar groups, as well as in almost every new book that was published on solar energy. One reason for our distress was that the interior shots featured special furniture which had been brought in by the photographer—his own props which he used instead of our furnishings. When the photographs kept reappearing we became concerned that at some point they might be used in connection with some group or program which we did not support, possibly a group to which we were fundamentally opposed.

After some investigation we learned that it is fairly standard procedure for house photographers to use props or special furnishings unfamiliar to the owner. We also found that most professional photographers copyright their work, in the same way that authors do, to protect their property and therefore their income. In this way their photographs become their own exclusive possessions and they have the right to use them when and where they see fit.

But this was our only negative experience with the many photographers who came to Sol-A-Sea. We enjoyed having most of the other young men who arrived with their cumbersome picture-taking equipment. They were usually very affable and sincere environmentalists with a special interest in solar energy. We were particularly fond of a young man who came all the way from California to take pictures of Sol-A-Sea for a new book. He arrived at nine a.m. and was still with us at eleven p.m. We enjoyed his company, too!

Soon after this visit, a motion-picture photographer from the Washington, D.C. area arrived to shoot scenes for a documentary being prepared by the government. The film, entitled *The Solar Generation*, was aimed at schoolchildren and featured small-scale applications of solar power.

A few weeks later, we received a call from Los Angeles with the voice on the other end saying, "Jacques Cousteau calling." I was sure the call was someone's idea of a practical joke. Although we lived on the water and loved to swim, we did not own wet suits and did not contemplate any undersea explorations. Why would this internationally known marine researcher be singling us out?

As the telephone voice continued, I realized I wasn't actually talking with Cousteau but with a member of his executive staff. The Cousteau Society, the voice informed us, was producing a television series on environmental problems entitled *Oasis in Space*. They hoped to include Sol-A-Sea in the segment on energy alternatives.

C.B. and I were delighted with the idea and realized that this film would reach a potential audience of millions. It would be a very effective way of helping solar energy make its mark on the public. An appointment was set up for the filming some two months hence.

When the long-awaited day finally arrived, we were ready for our distinguished visitors soon after daybreak. I had cleaned and polished everything in sight and sported a new coiffure for my television debut as a solar homeowner.

The morning dragged on. Just before the appointed hour, the jarring sound of the telephone intruded on our fantasies. This time a clear, clipped feminine voice said, "So sorry, but some of our equipment broke down and we can't seem to get it fixed here. We're just going to have to turn around and go back to California. We'll keep in touch. We'll see you another time." So they were off to California and we were high and dry and all prettied up for nothing.

I was crushed. We'd told many of our friends of our imminent debut. We'd just have to sit tight until the Cousteau crew could get it all together again. It did cross my mind fleetingly that they might just never come back. Such ruminations were dispelled in less than a month when we received another call from Los Angeles and another date was set for the taping. This time I couldn't take the whole day off from work and C.B. could not be present. I returned home around noon only to find that the Cousteau crew had arrived hours ahead of schedule. The place was swarming with cars and vans and buzzing with the babble of French.

The charming and unassuming Cousteau producer greeted me and suggested that I take him on a tour of the house while all the equipment was being set up. During the tour he questioned me about our background and our experience living in Sol-A-Sea. It didn't occur to me at the time that this was in fact a dress rehearsal for the formal interview to follow. It did serve as a good warm-up to prepare me for fielding his questions later on camera.

The actual filming took about an hour. After a brief social visit, the whole crew packed up and loaded their gear. Within minutes the entire entourage had vanished to continue their cross-country travels.

For weeks afterward we faithfully watched the Cousteau series on television. As time went by we began to feel somewhat shamefaced. We'd told so many people about the upcoming program and we were beginning to suspect it would never be released. Our worst suspicions were confirmed when the now-familiar Cousteau stationary arrived a few weeks later. This time the letter was an apology from the executive producer, who explained that they had dropped the sequence on solar homes. However, they were holding the footage "in abeyance with a possibility that it may be incorporated into some future production." If it ever was, we missed seeing it.

After experiencing all this disappointment over the Cousteau program, it is ironic that I actually slept through what could have been my international television debut. One fall day about a year after we'd moved into Sol-A-Sea, I was nursing a bad cold and slept away the afternoon. When I awoke, I found C.B. looking smug and amused. With a little smirk he announced that I had not only missed meeting some interesting visitors, but had also lost an opportunity to appear on television. While I'd slept, eight Japanese engineers and their interpreter had arrived and taken footage to be shown the following week in Tokyo. These Oriental gentlemen rounded out our geographic record: Asia was the last of the five continents to send us a delegation.

While our visitors were often unique, so was some of our publicity. Our grandson called excitedly about an aerial view of the house that he had come across in a magazine he was using for a school project. At first, he hadn't been sure it was our house; the picture was so unusual. On closer inspection he recognized our deck furniture and raced to the phone. He wondered why we hadn't told him and his parents about the article. But we knew nothing about it ourselves and had no idea how or when the picture had been taken.

Sol-A-Sea did receive other excellent publicity in our national press. One newspaper article that was widely read appeared in the *New York Times Magazine* on Sunday, April 7, 1974. It was written by the environmentalist author Wilson Clark, whose book, *Energy for Survival*, was about to be published.

The *Times Magazine* carried three full pages of material and included fine photographs of our unfinished house. There were close-ups of the collectors and the storage tank and a simplified diagram of our solar heating system. Many significant, well-documented facts about solar energy were brought out, and solar heating was stressed as an immediate and practical alternative to conventional fuels. According to Clark, government research proved that solar heating was already cheaper than electric heat in many parts of the country. He quoted a well-known firm of consultants, Arthur D. Little of Cambridge, Massachusetts, who predicted that by 1984 the market for solar products would reach $1 billion.

Clark assured readers the technology for solar energy was simple and already well developed and, as one engineer put it, required the most "trivial" kind of technology. Should solar energy provide only 1 percent of the nation's energy needs by 1983, it was estimated that the U.S. would save approximately 100 million barrels of oil every year.

Clark also introduced the concept of life-cycle costing in relation to the homeowner's investment in solar equipment. By calculating the probable savings in fuel bills over the full life cycle of a house (about forty years), Clark showed how homeowners would be more than compensated for their initial capital investments in solar equipment.

Sol-A-Sea's visitors and fans knew no age limits. During the first two or three years our guests included everyone from the tiniest infant in arms to some distinguished gentlemen well on in years.

Two of the latter arrived late one Saturday morning. We'd received one of Ev's familiar telephone calls earlier in the day just as we were preparing for luncheon guests. Ev said he was in the company of two fine gentlemen who were extremely interested in solar heating and who would very much like to visit Sol-A-Sea.

It was difficult to turn Ev down, although we were rushing and already far behind our schedule. He agreed to hold off these visitors for at least another hour. Just before hanging up, Ev added, "Oh, by the way, these two men are in New Haven for their reunion. They're the two oldest living graduates of Yale!"

Very shortly, this well-preserved, spunky pair of nonagenarians were at our door. The taller and more portly of the two sported a walrus mustache and seemed to be leading his more reserved and reluctant friend by the hand. This tall gentleman was as sprightly and spry as the average forty-year-old and as charming, ardent, and loquacious a solar enthusiast as we ever hoped to meet. Although his retiring companion seemed somewhat embarrassed by his classmate's exuberance, we found this fellow's antics charming and beguiling.

Before long the more aggressive member of the team was hopping across our roof, examining the collectors at close range. Later we found him scampering about the beach taking shots of Sol-A-Sea with sophisticated photographic equipment.

After a while, C.B. persuaded this vigorous guy to pause long enough in his explorations to join us in a glass of wine. Our friend seemed to be relishing every minute of his visit and we were equally enthralled with him. He told us at great length about his engineering background and how he had established his own research foundation for the study of natural forms of energy at one of the leading universities on the West Coast. After Yale's reunion, he planned to attend some meeting overseas in connection with his current survey of solar developments.

Long after our luncheon guests arrived, our new friend showed no sign of leaving—although his silent partner had been edging toward the door for some time.

Finally, this game gentleman rose to his full height as he prepared to leave. Then, taking several giant strides across the room, he suddenly leaned over and planted a huge kiss squarely on my astonished lips—walrus mustache and all! While I was recovering from this unexpected demonstration, he hastened to explain, "You know I'm going on ninety-four; I may never see you again!"

Our admirable and winsome visitor later wrote to us, saying he was recommending "Mr. Barber's service to the Energy Task Force authorized by Congress to find means of cutting down the heating costs of public educational buildings." He added that the fuel bill on the Yale campus was over $8 million annually and still climbing. He was impatient with the authorities who were hesitant about putting solar panels on the roofs of the university buildings. The panels, he wrote, "would not injure them in the least and would save half their fuel bill from now on." He closed by saying, "Our visit with you was the highlight of my trip."

While our Yale friend startled us with his unusual leave-taking, another guest surprised us with his dramatic, spontaneous entrance. This young man sprang full-blown from the sea! Attracted to Sol-A-Sea by the sun's bright reflection off our solar panels, he paddled his canoe up to our beach and presented himself on the deck in dripping wet bathing trunks.

Gradually Sol-A-Sea has become less alluring. When we moved into our home in 1974, it was perhaps one of twenty solar buildings in the entire country and was, we believe, unique on the East Coast. But two years later there were probably over a hundred solar buildings throughout the nation. Sol-A-Sea was no longer such an oddity or star attraction.

After 1976 our visitors tapered off noticeably. Now those who came to see Sol-A-Sea seemed far more sophisticated and better informed. They were more interested in the efficiency of our solar operation than in the details of construction or the individual solar parts. They communicated their enthusiasm and contributed greatly to our solar knowledge and understanding of what was going on in many different parts of the country and in other parts of the world. We always felt we were gaining more from them than we were able to contribute.

The crowning glory for Sol-A-Sea and for Don, too, was an honor bestowed by the Connecticut Society of Architects in 1976 —a special "citation of merit" to Don for his "design of a three-bedroom residence that will obtain approximately 60 percent of its annual space and water heating requirements from a modular flat-plate solar collector system." The citation gives special praise to several of Sol-A-Sea's energy-saving features: the pocket doors, which permit closing off large spaces; the clerestory windows, which maximize our natural daylight and ventilation; and the broad overhangs, which reduce heat gain during the hot summer months. The award was presented in the hope that "the innovations employed in their design will become prototypes for future projects."

And in a very real sense Sol-A-Sea *has* become a prototype for solar homes. What began as our private effort to build a seaside retirement home has become a symbol of a solar-heated home.

Photo by Robert Perron

9

Solar Living

Springtime at the shore is a time for renewal and reawakening as it is elsewhere. It is when flowers and grasses grasp eagerly at the sand, clawing their way across the wide beach; when tiny plovers beep shrilly to divert us from their shallow nests among the dunes; and it is the season when horseshoe crabs—those ancient armored creatures—come plodding up the sand, carefully depositing eggs to carry on the species.

By summer there are other sights and sounds: the rustling of the incoming tide as it washes in to shore, its waves breaking gently on the sand, the flapping of canvas and nylon as the huge sails climb up the masts and suddenly balloon as they fill up with wind, the steady muffled roar of the cabin cruisers, and the noisy whirring of outboards.

Photo by Robert Perron

Late summer and early fall often bring fierce tropical storms, leaving destruction and desolation along the coastline. Then the Sound becomes a seething maelstrom, its swift currents and mighty swells forcing giant waves to crash onto the beach. In August, 1976, when Belle was battering the Atlantic shores, we were evacuated from Pilots Point and we fled to higher ground along with dozens of other shoreline refugees. We huddled together in a nearby motel. All night long we watched the trees waving madly and listened to the howling of the wind. In the morning we returned to the Point, carefully picking our way over fallen limbs and wires. What a relief to find Sol-A-Sea damp and droopy but unscathed! Not a single collector had been cracked or marred. Only one bright strip of flashing, dangling in the wet breeze, gave evidence of the hurricane's fury. Everything inside was also intact and there wasn't a leak or a puddle to be found anywhere. Nothing had been moved or disturbed.

Fortunately, the eye of the storm had not coincided with high tide and so did not produce the expected fifteen-foot tide. Nevertheless, the water did rise many feet above normal and there was considerable flooding and property damage in our area.

Winds of sixty miles an hour aren't unusual on the Point. Their whistling and wailing penetrate even our heavy insulation. Often these high winds accompany the big Northeasters which bring heavy rain or deep snow. But even in the harsh and bitter winter of 1977-1978, our sturdy collectors held up under a snowfall of more than two feet—no small tribute to Ev's skills and competence.

Though our winters are cold and wet along the shore, compared with most inland locations they are surprisingly temperate. A three- or four-inch snowfall, in wooded areas five or ten miles north of us, often produces only a light dusting of snow at the beach. But in one recent severe winter, the harbor froze solid clear out to Menunketesuck, a tiny island immediately off Pilots Point. Even the lobster trawlers and the big draggers were immobilized for days, while the wintry tableau was iced into stillness. As soon as the weather moderated, huge ice floes came crashing together, knocking off big chunks until most of the ice melted and disappeared. Within a day or two, the familiar fishing boats returned, slowly and stealthily picking their course as they resumed their appointed rounds.

No matter what the season, the end of the day at the shore often brings spectacular, breathtaking sunsets. With sky and water as a backdrop, the shades of mauve, pink, and gold, first so subtle and then so intense, hardly seem real. As these astonishing tones merge and melt, flow and dilute into paler shades, the fiery sun overtakes them, dragging them and their beautiful mirrored reflections down into the watery horizon.

In addition to enriching our lives aesthetically, Sol-A-Sea has brought us many practical benefits. We soon found there was no need to extend our solar system to include air conditioning, although it had been designed for easy adaptation. The very same solar ducts and vents used for heating could also be used for cooling and could be made operational with almost no additional expense. But there has seldom been a time when our sea breezes haven't kept us comfortable. Sometimes in midsummer we do seek refuge from the sun's searing heat under our large deck umbrella; however, the heat inside the house has never become unbearable.

Granted, Sol-A-Sea has brought us comfort and many new satisfactions, but exactly what did Sol-A-Sea do for our budget?

Paying for the house in gradually increasing installments over the course of a year made it less painful to expend much more than we had originally planned. All told, we paid Dick Riggio $66,969, or more than 10 percent over our planned budget of $60,000. Of the increases, less than $700 was for charges related to the solar features.

In addition to Dick's bill, large outlays went for a built-in vacuum system, garbage disposal, electric garage door, and built-in television wiring and antenna—totaling around $2,000. Another $2,000 went for carpeting the living room, bedrooms, and halls; an additional $9,000 was divided up among such items as Roman shades, blinds, furniture, and upholstery; and a final "umbrella" entry of $1,850 covered track lights, chandeliers, outdoor fixtures, wallpaper, deck furniture, and other items.

When we added the $15,000 total for these items to Dick's bill of $66,969, we had expended $82,000—or more than twice what we had originally planned to spend. And even that wasn't the whole story. If we added Don's fee for architectural services, Ev's fee for designing and installing the heating system, legal costs, and the $20,000 for the land, our total outlay came to well over $100,000.

C.B. had said right along that he did not want to know how much Sol-A-Sea cost, because if he did he might never be able to enjoy it. Although $100,000 was a tidy sum for us to pay in 1974—and far more than we'd ever dreamed of paying—five years later, in 1979, a $100,000 house isn't necessarily in a luxury class, or even considered expensive or unusual.

Even with our greatly expanded budget, no one on our solar team got rich, or even made much money. All three of them considered themselves lucky if they broke even. Don's fee remained fixed despite our increases. His only extra charge was for redrawing and reprinting the later revisions of the plans, an added sum of $600.

Ev was probably not fortunate enough to break even on his investment of time and energy. The $7,000 we paid him for the mechanical system might have covered the cost of materials but couldn't possibly have compensated him for the countless extra hours he spent installing, adjusting, improving, and servicing our solar system. And the record should clearly show that Ev paid back every penny C.B. had loaned him, with interest.

Nor could Dick riggio have found his contract for building Sol-A-Sea especially lucrative. The increase in the cost of materials and labor during construction must have whittled down his normal 10 percent profit considerably. And Dick also contributed many extra hours to fulfill all the house's special requirements.

Despite our inflated budget, Sol-A-Sea has been a bargain for us in many ways. The solar heating system has proved a wise investment and there's little doubt that we could make a handsome profit if we were to put the house on the market today. A far greater profit than we would realize from a conventional home. The solar features have become progressively more valuable, increasing our equity in the house and paying for themselves many times over.

Before we could estimate our exact fuel savings in dollars and cents, we needed some additional input about the actual operation of our solar heating. Although our solar system was initially designed to provide 45 to 55 percent of our heating needs, in practice it has supplied over 60 percent.

Many different factors have combined to boost the efficiency of our solar operation. Not the least of these is the unusually large amount of passive solar heat collected by the house itself. With our large glass areas on the south side of the house, Sol-A-Sea can utilize a great deal of available solar energy without employing any mechanical devices. The house's compactness of design, its tightness of construction, and the thoughtful placement of each window and door have all contributed to our favorable solar performance.

And there was another bonus from our seaside location: the strong reflection of the sun on miles of unobstructed water and on the wide, sandy beach on the southside have appreciably increased the amount of solar radiation available to us. Although Ev had known that radiation would be increased by reflection on the first row of collectors, it was not anticipated that the second and third row of collectors would also benefit from reflected radiation.

We must bow to all three of our solar experts: Don for his pleasing and efficient plans and layout, Ev for his engineering genius in developing our solar heating system, and Dick for his special abilities and skills in carrying out all the building details so expertly. Each contributed his own special talents, and the combination had somehow produced the magic to make our solar home a success.

And now to get down to the nitty-gritty: How much cold, hard cash has our solar heat actually saved us? Ev estimated that during 1974-1975, our first winter at Sol-A-Sea, our solar heating had supplied between 60 and 65 percent of our heating needs. During that first heating season, between October 1974 and May 1975, we burned 670 gallons of heating oil, far less than the 1500 gallons a house the size of Sol-A-Sea would be expected to use. In other words, we used 830 gallons less than we would have expected to use with a conventional oil heating system. Since the average annual cost of oil in that period was $.397 per gallon, we saved approximately $329.51 during our first heating season.

During our second winter we used 750 gallons of oil, which had now jumped to $.427 per gallon. At this new price, we estimated that we had saved $320.83 that year. The following year, 1976-1977, we used 707 gallons, but the average cost of oil had now jumped another 4 cents to $.466 and our estimated savings in fuel amounted to $369.54. And during the fourth winter, 1977-1978, we used 769 gallons of oil selling at $.495 per gallon. Our savings that year we estimated at a total of $361.84.

In other words, in each of our first four heating seasons, we believe we paid less than half of what we would have paid to heat our house entirely with oil. In those four years alone, based on these calculations, we had realized a total savings of $1,391.30, close to 30 percent of the $3,500 we'd invested in solar equipment.

Other factors to be considered in estimating our savings in fuel are our winter vacations. Although we took vacations of only a week or two during the first two years in Sol-A-Sea, after that we began to extend our vacations by several weeks. During our absence, the thermostat was set at 60°. This too tended to reduce the amount of oil needed from the backup system when the solar heating couldn't maintain this temperature level.

Another variable to be considered is the severity of the winters. Since this would fluctuate from year to year, the estimated annual 1,500-gallon heating requirement for our house would also fluctuate.

Another way of estimating solar savings is by figuring life-cycle costing, the total savings in heating costs and maintenance over the full lifetime of a house, or over about forty years. The savings are computed by comparing the actual cost of heating our house with the estimated cost of heating a non-solar house during the same period. The cost of installing solar equipment in addition to the backup system would also need to be included. This is often a more realistic way of calculating actual long-term savings which can be made by using solar heating methods.

While $3,500 for a solar heating system was a real bargain in 1973, today it would be a giveaway. And of course the additional $3,500 for the auxiliary system would also be unrealistic now. Today, a combined solar and auxiliary heating system is apt to cost between $10,000 and $18,000. Increased labor costs as well as increases in the cost of materials have contributed to higher prices for all heating systems, including solar.

Ev estimated that our collectors cost approximately $8 per square foot in 1973. At that time it was generally believed the cost of collectors and other solar equipment would drop considerably as soon as they were mass-produced and could be sold competitively on the open market. However, inflation and other factors seem to have destroyed this expectation. Actually, collectors cost about $12 per square foot today, or 50 percent more than they did in 1973. Whereas our 2,000-gallon steel storage tank cost $500 in 1973, today it would probably cost close to $900. And while our auxiliary oil-fired heater cost $450 when it was installed, today it would probably cost close to $750. Our hot air distribution system (blowers and heat exchangers) which Ev priced at about $800 in 1973, would probably sell for about $1,200 today.

Our solar heating system is no more difficult to operate than conventional heating systems. It requires no more effort than setting a thermostat. We just choose the "comfort zone" temperature we want to achieve and set the thermostat accordingly. Of course there is one difference; that is, if the solar heat already collected and stored is insufficient to provide the required temperature, the oil-fired auxiliary system goes into action. But all this is entirely automatic.

There is only one way we can be sure our heat is coming totally from our solar system; that is, by shutting off the switch marked "auxiliary circulator." Sometimes we shut the circulator off to see how long we can go without supplementary heat. Unless the outside temperature is very low, we can continue to be comfortable for at least a day or two in moderate winter weather even when there is little sun. The original predictions that we could manage on solar heat for three days without any sunshine seem to have been overoptimistic.

As with any experimental system, changes and adjustments have been necessary. Most of these have been minor but we did run into a big problem with our original solar domestic hot water heating, which was actually in operation less than a year (from October 1974 to April 1975). At that time Ev had to add an anticorrosive chemical, sodium chromate, to the water in the storage tank to inhibit the tank's deterioration. And it was conceivable that this toxic chemical could leak into our domestic water and contaminate it.

After evaluating our whole operation, Ev decided to separate our space heating and water heating operations.

Our new and separate hot water unit was installed recently and it is too early to assess its efficiency with any accuracy. But we do know that solar hot water heating is economically feasible today and holds great promise for the future. Since the heating of water sometimes uses as much as one-third of a home's total heat output, solar hot water systems could result in significant savings for millions of householders.

For those who already own their home, adding a small solar hot water heating unit offers an economical and relatively uncomplicated way to reduce fuel bills. In warm climates, simple solar water heaters often use solar energy directly—that is, the sun simply heats up the water in a container directly exposed to it. They can sometimes be installed for as little as $600. In cooler climates, solar water heaters are usually more complicated and more expensive because an antifreeze solution must be used in the collectors if there is any possibility of freezing temperatures. This requires the installation of a pump and a heat exchanger, which adds considerably to the cost.

Complete solar hot water systems today cost between $1,900 and $3,500, but the payback period gets shorter all the time as gas, oil, and electric heating become more costly.

Other matters needed adjustment after we moved into Sol-A-Sea. Despite our earlier doubts and questions, our solar heating was set up with just one heating zone and just one thermostat to regulate the heat throughout the house. We had grown accustomed to three-zone heating in our Stamford home and had anticipated a similar arrangement at Sol-A-Sea. But Ev did not think multiple zoning was necessary with our particular hot-air heating system, which is supposed to produce a complete change of air in the house every ten minutes. He believed that this constant circulation of heated air would distribute the heat evenly throughout the house.

This has not proved to be so, mainly because of the unusual amount of passive solar heat gain on the south side of the house. It isn't unusual for this exposure to heat up into the high seventies or low eighties during a bright winter afternoon. When this happens, our one and only thermostat, located near the huge south-facing windows, turns off the heat in the entire house. This means that rooms in other parts of the house—particularly those with northern exposure—are somewhat chilly.

Ev has been able to alleviate this problem to a great extent by installing an additional fan. This circulates the warm air through the house while the main heating system is off. The fan operates separately from the rest of the system and must be turned on manually. It is specifically designed to take the heat from the warmer side of the house and blow it through to the colder areas, thus equalizing the temperatures throughout the house.

While the fan is an adequate solution, if we had it to do over again, we would still opt for multiple-zoned heating. It is a lot easier to push a thermostat up or down than to keep opening and closing heating registers and pocket doors or to continue turning on an extra fan, and it might very well conserve energy too.

Nor have we yet become complete devotees of hot-air heating. While this kind of heat is clean, it does tend to be noisy. It took us some time to get used to the frequently alternating on-and-off operation of the blower. When we first moved in, the large fan made such a loud roar it sometimes woke us up during the night. Ev was able to solve this problem to our satisfaction by lining some of the return ducts with coated fiberglass. This application slowed down the fan and muffled the noise and we are no longer conscious of the operation of the blower.

We consider ourselves lucky to have built our solar home when we did. An increasing number of people are finding it very costly to build a single-family dwelling of any kind. But it isn't necessary for solar enthusiasts to invest in an entirely new house. There is such a thing as "retrofitting"—applying solar equipment to an existing building. Retrofitting a house with solar fixtures can be an economical alternative to building, and there are many ways it can be accomplished.

In most cases it is feasible to add at least some kind of solar heat to an existing house. But to be well suited for retrofitting, a house should meet certain basic requirements. One is that sufficient solar radiation must be available to make solar heating a wise investment. Parts of this country and other areas in the world are almost continuously under a cloud cover. In other places, the air is so polluted as to make it impossible to collect the solar radiation necessary for either solar space or hot water heating. Second, the house must either face south or have a south-facing roof, wall, or nearby area where solar panels can be installed. The area designated for the collectors must have complete and unobstructed access to a maximum amount of direct sunshine. There cannot be any nearby trees or high buildings which could conceivably interfere with the available sunshine. Finally, the house should be as tight as possible. All windows and doors must fit snugly and be well caulked and weather-stripped. The side walls, roof, and flooring must have extra-heavy insulation to keep heat loss to a minimum.

Many existing homes can meet these requirements and would be excellent candidates for retrofitting. But unless all three primary qualifications are met, the addition of solar equipment would probably be less than satisfactory and would provide a questionable return on the capital invested in it.

Much has happened since the day we took those first faltering footsteps to find a builder in 1972. The frustration, anxiety and rancor we experienced while the house was under construction have been shed long since, discarded along with the bent nails and piles of sawdust. In their place we have found the satisfaction of having helped to create a cheerful, comfortable and beautiful energy-conserving home, and of being able to live in it and enjoy it.

And now, there is a lot of living to be done at Sol-A-Sea.

Epilogue

We must look to the upcoming solar generation, our children, and our children's children to explore, develop, and expand many of the methods and uses to which our greatest resource can be applied. Using the sun's energy to heat homes may become as ordinary and familiar to the rising generation as heating with gas, oil, or coal has been for us. We must remind ourselves that this is the generation that has grown up knowing that men have walked on the moon, while most of us grew up believing there was a man in the moon.

Solar heating and solar homes will grow and spread and their methods and techniques will become more sophisticated. Before long, solar architecture and design should become an accepted part of our national scene. Solar heating is still in its infancy. It is hoped that posterity can look forward to many improvements and refinements in solar heating, much as the steam engine has been adapted and modified over the years.

The long and arduous journey into the Solar Age has already begun. And those who have taken the first step must make certain that those who follow will work toward maintaining our way of life while continuing our efforts to preserve and protect our environment.

One important way to reach these goals is by developing the potential of the greatest of all our resources — the sun.

Appendixes

1. Federal Agencies

Department of Energy (DOE)

The Department of Energy (DOE) came into existence on August 4, 1977 and took over prime responsibility for research, development, regulation and administration of all energy resources. James R. Schlesinger, the first Secretary of Energy, is the twelfth member of the President's Cabinet.

Department of Energy—Regional Offices

Region I:	150 Causeway Street, Room 700, Boston, MA 02114 (617) 223-5257 (CT,ME,MA,NH,RI,VT)
Region II:	26 Federal Plaza, Room 3206, New York, NY 01007 (212) 264-0560 (NJ,NY,PR,VI)
Region III:	1421 Cherry Street, Philadelphia, PA 19102 (215) 597-0792 (DE,DC,MD,PA,VA,WV)
Region IV:	1655 Peachtree Street, N.E., Atlanta, GA 30309 (404) 881-2062 (AL,FL,GA,KY,MS,NC,SC,TN)
Region V:	175 West Jackson Boulevard, Chicago, IL 60604 (312) 353-5779 (IL,IN,MI,MN,OH,WI)
Region VI:	P. O. Box 35228-2626 West Mockingbird Lane, Dallas, TX 75235 (214) 749-7621 (AR,LA,NM,OK,TX)

Region VII: 324 East 11th Street, Kansas City, MO 64106 (816) 374-2061 (IA,KS,MO,NB)

Region VIII: P. O. Box 26247, Belmar Branch, 1075 South Yukon Street, Lakewood, CO 80226 (303) 234-2420 (CO,MT,ND,SD,UT,WY)

Region IX: 111 Pine Street, San Francisco, CA 94111 (414) 556-0418, 7157 (AZ,CA,HI,NV,Guam,Pacific Trust Territories)

Region X: 1992 Federal Bldg., 915 Second Avenue, Seattle, WA 98174 (206) 442-7285 (AK,ID,OR,WA)

Department of Energy Report Collections:

Listed below are the libraries that purchase and maintain microfiche collections of Department of Energy publications. Most of the libraries have microfiche reader-printers or photocopy centers which reproduce enlarged copies from microfiche. Charges for reproduction services vary. Ask the librarian for help in using the *Energy Research Abstracts* or in finding particular publications.

Alabama

Auburn University,
Auburn, AL 36830
Tuskegee Institute,
Tuskegee, AL 36088

Arizona

University of Arizona,
Tucson, AZ 85717

California

University of California,
Davis, CA 95616
University of California,
Los Angeles, CA 90007
University of California,
Santa Barbara, CA 93107

Colorado

University of Colorado,
Boulder, CO 80302

District of Columbia

Library of Congress,
Washington, DC 20540

Florida

University of Florida,
Gainesville, FL 32601

Georgia

Georgia Institute of Technology,
Atlanta, GA 30332

Hawaii

University of Hawaii,
Honolulu, HI 96825

Idaho

Idaho State University,
Pocatello, ID 83201

Illinois

University of Illinois,
Urbana, IL 61801

Indiana

Purdue University,
Lafayette, IN 47901

Iowa

Iowa State University,
Ames, IA 50010

Kansas

Kansas State University,
Manhattan, KS 66502

Kentucky

University of Kentucky,
Lexington, KY 40506

Maryland

Johns Hopkins University,
Baltimore, MD 21218
University of Maryland,
College Park, MD 20746

Massachusetts

Massachusetts Institute of
Technology, Cambridge,
MA 02138
Worcester Polytechnic Institute,
Worcester, MA 01601

Michigan

University of Michigan,
Ann Arbor, MI 48106

Mississippi

Mississippi State University,
State College, MS 36762

Missouri

University of Missouri,
Columbia, MO 65201
Linda Hall Library,
Kansas City, MO 66101

New Jersey

Princeton University,
Princeton, NJ 08540

New Mexico

University of New Mexico,
Albuquerque, NM 87131

New York

New York State Library,
Albany, NY 12224
State University at Albany,
Albany, NY 12222
Cornell University,
Ithaca, NY 14850
Columbia University,
New York, NY 10025
University of Rochester,
Rochester, NY 14627
Syracuse University,
Syracuse, NY 13210

North Carolina

North Carolina State University,
Raleigh, NC 27607

Ohio

University of Cincinnati,
Cincinnati, OH 45221
Cleveland Public Library,
Cleveland, OH
University of Toledo,
Toledo, OH 43606

Oklahoma

University of Oklahoma,
Norman, OK 73069

Pennsylvania

Carnegie Library,
Pittsburgh, PA 15213
Pennsylvania State University,
University Park, PA 16802

Puerto Rico

University of Puerto Rico,
San Juan, PR 00931

South Carolina

University of South Carolina, Columbia, SC 29208

Texas

University of Texas, Austin, TX 78712

Texas A&M University, College Station, TX 77846

Rice University, Houston, TX 77005

Utah

University of Utah, Salt Lake City, UT 84601

Virginia

University of Virginia, Charlottesville, VA 22903

Washington

Washington State University, Pullman, WA 99163

University of Washington, Seattle, WA 98105

West Virginia

West Virginia University Morgantown, WV 26505

Wisconsin

University of Wisconsin, Madison, WI 53715

Housing and Urban Development

The Department of Housing and Urban Development (HUD) is charged with administering all federal solar energy loans and conservation grants for homeowners. Write to: HUD, Washington, D.C. 20410.

The Solar Energy Research Institute

SERI was created by Congress through the Solar Energy Research, Development, and Demonstration Act of 1974. A contract was awarded to Midwest Research Institute in April 1977.

SERI is mission-oriented and is dedicated to supporting, at the earliest feasible time, those applications of solar energy that are commercially attractive and environmentally acceptable. This technology will contribute to the establishment of a solar energy industrial base that will foster widespread usage, thus creating a major alternative source of energy for the nation.

One of SERI's objectives is to develop and implement the Solar Energy Information Data Bank (SEIDB), a highly advanced resource that will assure that those individuals who are making decisions concerning the solar option receive needed information and data in an understandable and usable format in a prompt and efficient manner. SEIDB is located at the Solar Energy Research Institute, 1536 Cole Boulevard, Golden, CO 80401. Phone (303) 234-7171. Regional offices are: The Northeast Center, 70 Memorial Drive, Cambridge, MA 02142. The Southeast Center, Governor's Energy Advisory Council, 7703 N. Lamar, Austin, TX 78752.

Superintendent of Documents
Government Printing Office (GPO)
Washington, DC 20402

Holdings: Books and technical reports published for the federal government by the Government Printing Office (GPO).

How to Order: Write to the GPO. Send for the free *Solar Energy Bibliography.* GPO bookstores, located in major cities, have popular titles. Regional GPO depository libraries (listed in the "Where to Read" section) have one copy of most GPO publications available for reading on-site free.

Typical Publications: General interest books, maps, pamphlets, periodicals. Selected U.S. Government Publications (lists 150-200 popular publications for sale, ten times per year); and Solar Energy (Bibliography No. 9). *Solar Heating and Cooling; An Economic Assessment,* NS1, 2:SO 4/14; *Solar Dwelling Design Concepts,* HH1, 2:SO 4/3; *Energy Research and Development Inventory* (5 volumes), #Y4, SC12:94-2/u; Monthly Catalog of U.S. government publications; many more.

Consumer Information Center

U.S. Public Documents Distribution Center
PO Box 713
Pueblo, CO 81002

How to Order: Write to the above address. Send for free issue of *The Consumer Information Catalog.*

Areas of Interest: Since 1970 has published the quarterly Consumer Information Catalog (free), lists 200 consumer publications.

Typical Listings: *Buying Solar,* 114E, $1.85; *Solar Energy and Your Home,* 646E, free; *How to Get a Patent,* 126E. 75¢.

National Climatic Center

Parent Organization:
U.S. National Oceanic & Atmospheric Administration Environmental Data Service
Federal Building
Asheville, NC 28801
(704) 258-2850

Areas of Interest: Weather-related data including wind direction, wind speed, solar radiation.

Holdings: Past weather records; 10,000 books; 94,000 pamphlets.

Information Services: Answers brief inquiries, library open to public (free).

National Solar Heating and Cooling Information Center

Parent Organization: Franklin Institute Research Laboratories
PO Box 1607
Rockville, MD 20850
(800) 523-2929
(800) 462-4983 (PA)
(800) 523-4700 (Alaska and Hawaii)

Areas of Interest: Solar thermal energy for buildings, including collectors, storage, materials, heating and cooling systems, water heating, electric power.

Holdings: Solar energy applications information including articles, books, bibliographies.

Publications: Data compilations, bibliographies, technical reports, pamphlets, fact sheets (many free).

Information Services: Answers general questions for the public, provides information on research-in-progress, makes referrals to other sources of information (free).

National Technical Information Service
5285 Port Royal Rd.
Springfield, VA 22150

This agency processes over 20,000 pieces of information daily and can supply most government reports on energy. It issues a semiannual bibliography on solar energy and a monthly journal on new publications.

2. Solar Legislation and Financial Incentives by State

Within the past several years, most states have passed legislation relating to solar energy and its use, particularly laws providing financial incentives for solar consumers. The following tables are divided into four areas: Section A — Financial Incentives; Section B — Grants and Loans; Section C — Land Use Provisions; and Section D — Standards and Regulation of Construction. Updated as of January, 1979, this listing is available from the National Solar Heating and Cooling Information Center, P.O. Box 1607, Rockville, MD 20850.

A. Financial Incentives

State	Law	Description	Contact
Alaska	Ch. 94 Laws of 1977	Alaska allows a 10% residential fuel conservation credit of up to $200 per individual or married couple for money spent on the following: 1) insulation; 2) insulating windows; 3) labor related to items 1 and/or 2; 4) alternate energy systems which are not dependent on fossil fuel, including solar, wind, tidal, and geothermal. Expires 12/31/82.	State Department of Revenue Income Tax Division Pouch SA State Office Building Juneau, AK 99811 (907) 465-2326
Arizona	Ch. 93 Laws of 1975 as amended by Ch. 129 Laws of 1976, as amended by Ch. 112 Laws of 1978	When solar energy devices are added to business or investment property, the taxpayer may elect to depreciate, to amortize over 36 months, or to claim an income tax credit. This credit, during 1979, may equal 30% of the cost of the solar energy system. It is reduced 5% each year until the law expires 12/31/84. The maximum credit is $1000. Credit is available to individual taxpayers at the same rate.	State Department of Revenue Box 29002 Phoenix, AZ 85038 (602) 255-3381
	Ch. 165 Laws of 1974	Solar energy devices are exempt from property taxes through 12/31/84.	

	Ch. 42 Laws of 1977	Solar energy devices are exempt from Transaction Privilege and Use Taxes.	
	Ch. 112 Laws of 1978	A 25% credit is allowed for residential insulation and ventilation devices, such as insulating doors and windows. Maximum credit is $100. Credit expires 12/31/84.	
Arkansas	Act 535 Laws of 1977	Individuals may deduct the entire cost of solar heating and cooling equipment from their taxable income. The cost of other energy-conserving devices, such as insulation, storm windows, and motor-driven ventilation, may also be deducted.	State Department of Revenue Income Tax Section 7th & Wolfe Streets Little Rock, AR 72201 (501) 371-2193
California	Ch. 168 Laws of 1976 Ch. 1082 Laws of 1977 Ch. 1154 Laws of 1978	California provides personal income tax credit of 55% of the cost of a solar energy system, up to a maximum of $3000. If a system is installed in other than a single-family dwelling and the cost exceeds $6000, the credit equals 25% of the cost, or $3000, whichever is greater. In both single-family dwellings and other buildings, the cost of energy-conserving devices installed in conjunction with the solar energy system may also be included in the total cost used to calculate the tax credit. If a federal credit is claimed, the	Franchise Tax Board Attn: Correspondence Sacramento, CA 95807 (916) 355-0370

State	Law	Description	Contact
	Ch. 168 Laws of 1976 Ch. 1082 Laws of 1977 Ch. 1154 Laws of 1978	state credit is reduced by the amount of the federal credit. The same provisions apply to corporate taxpayers. The system must meet the criteria of the California Energy Commission. Eligible expenses for credit include attorney's fees, compensation, and recording fees associated with obtaining a solar easement. Credit expires 12/31/80.	Franchise Tax Board Attn: Correspondence Sacramento, CA 95807 (916) 355-0370
Colorado	Ch. 344 Laws of 1975	Solar heating and cooling devices are assessed for property tax at 5% of their value.	Local assessor or board of assessors
	Ch. 512 Laws of 1977	Individual taxpayers may deduct the cost of a solar energy device from taxable income. Corporate taxpayers may use the deduction in lieu of depreciation. Individuals and corporations may also deduct the cost of wind and geothermal energy systems.	State Department of Revenue Income Tax Division 1375 Sherman Street Denver, CO 80261 (303) 839-3781
Connecticut	PA 77-457	Solar collectors are exempt from sales tax through 10/1/82.	State Tax Department Audit Division 92 Farmington Avenue Hartford, CT 06115 (203) 566-2501

	PA 76-409 and 77-490	Municipalities are authorized to exempt solar heating and cooling or solar electrical systems from property tax for 15 years. Windmills and waterwheels are exempt. The amount of exemption equals the difference between an assessment including the solar energy system and an assessment including a conventional energy system. Installation must take place before 10/1/91. The taxpayer must apply to the local Board of Assessors. The system must meet the standards of the Commission of Planning and Energy Policy.	Local assessor or board of assessors
Delaware	Ch. 512 Laws of 1978	This law provides an income tax credit of $200 for solar energy devices designed to produce domestic hot water. Systems must meet HUD Intermediate Minimum Property Standards Supplement for Solar Heating and Domestic Hot Water Systems. Systems must be warranteed according to criteria set out in the law.	Division of Revenue State Office Building 820 French Street Wilmington, DE 19801 (302) 571-3360
Georgia	Act 1030 Laws of 1976, as amended by Act 1309 Laws of 1978	Real estate owners may claim a refund of sales tax paid for the purchase of solar equipment. Expires 7/1/86. Georgia constitution, Article VII, Section 1, paragraph IV. Expires 7/1/86.	State Department of Revenue Sales Tax Division 309 Trinity-Washington Building Atlanta, GA 30334 (404) 656-4065

State	Law	Description	Contact
	Georgia constitution, Article VII, Section 1, Paragraph IV.	Any county or municipality may exempt solar heating and cooling equipment and machinery used to manufacture solar equipment from property taxes. Expires 7/1/86.	Local city council or county board of supervisors
Hawaii	Act 189 Laws of 1976	A 10% income tax credit is provided to individuals and corporations who purchase solar energy devices that are placed in service by 12/31/81. The law also provides property tax exemptions for solar energy systems through 12/31/81. This exemption also applies to any non-nuclear and non-fossil fuel system and to any improvement that increases the efficiency of systems which use fossil fuel.	State Tax Department P.O. Box 259 Honolulu, HI 96809 (808) 548-3270
Idaho	Ch. 212 Laws of 1976	This law allows an income tax deduction for a solar heating/cooling or solar electrical system installed in the taxpayer's residence. The deduction equals 40% of the cost in the first year and 20% of the cost in each of the next 3 years; the maximum deduction in any year is $5000. This deduction also applies to systems	State Tax Commission 5257 Fairview Boise, ID 83722 (208) 384-3290

		fueled by wind, geothermal energy, wood, or wood products. Built-in fireplaces qualify if they have control doors, regulated draft, and heat exchangers that deliver heated air to substantial portions of the residence.	
Illinois	PA 79-943 (1975), as amended by PA 80-430 (1977)	A property owner who installs a solar or wind energy system may claim an alternate valuation for property taxes. The property is assessed twice: with the solar or wind energy system and also as though it were equipped with a conventional system. The lesser of the two assessments is used to compute the tax due. Owners must file a claim with the local Board of Assessors. Systems must meet the standards of the Division of Energy of the Department of Business and Economic Development.	Local assessor or board of assessors
Indiana	P.L. 15 (1974), as amended by P.L. 68 (1977)	The law permits the property owner who installs a solar heating and cooling system to have property assessment reduced by the difference between the assessment of the property with the system and the assessment of the property without the system. The owner must apply to the county auditor.	Local assessor or board of assessors

State	Law	Description	Contact
Iowa	Section 441.21, code of 1979	Installation of a solar energy system will not increase the assessed, actual, or taxable values of property for 1979-1985.	Local assessor or board of assessors
Kansas	Ch. 434 Laws of 1976, as amended by Ch. 346 Laws of 1977	The individual taxpayer is allowed an income tax credit of 25% of the cost of a residential solar energy system to a maximum of $1000. A solar energy installation on a business or investment property receives a credit equal to 25% of the system cost, $3000, or that year's tax bill, whichever is the least amount. The cost of an installation on a business or investment property can be amortized over 60 months. Wind energy systems are also covered by this law. Credit expires 7/1/83.	State Department of Revenue P.O. Box 692 Topeka, KS 66601 (913) 296-3909
	Ch. 345 Laws of 1977, as amended by Ch. 419 Laws of 1978	If a solar energy system supplies 70% of the energy for heating and cooling, the property owner may be reimbursed for 35% of his property tax for up to 5 consecutive years. Applies through 1985. Claims must be filed with the Department of Revenue.	

Louisiana	Act 591, 1978	Solar energy equipment installed in owner-occupied residential buildings or in swimming pools are exempt from property tax.	Local parish tax assessor
Maine	Ch. 542 Laws of 1977	Solar space or water heating systems are exempt from property tax for 5 years after installation. Eligible taxpayers must apply to the local Board of Assessors. Purchasers of solar energy systems may also receive a sales tax rebate from the Office of Energy Resources.	Local assessor or board of assessors (For Property Tax) Office of Energy Resources 55 Capitol Street Augusta, ME 04330 (207) 289-2196 (For sales tax)
Maryland	Ch. 509 Laws of 1975, as amended by Ch. 509 Laws of 1978	A solar energy unit will be assessed at no more than a conventional system needed to serve the building.	Local assessor or board of assessors
	Ch. 740 Laws of 1976	Baltimore City and any other city or county may offer property tax credits for the use of solar energy systems in any type of building. Credit may be applied over a 3-year period.	Local city or county department of revenue
Massachusetts	Ch. 734 Laws of 1975 as amended by Ch. 388 Laws of 1978	Solar energy systems are exempt from property tax for 20 years from the date of installation.	Local assessor or board of assessors

State	Law	Description	Contact
	Ch. 989, 1977	Sales of equipment for residential solar energy systems, wind power systems, or heat pumps are exempt from sales tax.	State Department of Corporations & Taxation 100 Cambridge Street Boston, MA 02204 (617) 727-4601
	Ch. 487, 1977	Corporations may deduct the cost of a solar or wind energy system from income. The system will also be exempt from tangible property tax.	State Department of Corporations & Taxation 100 Cambridge Street Boston, MA 02204 (617) 727-4201
Michigan	PA 132, 1976	Proceeds from sales of solar, wind, or water energy conversion devices used for heating, cooling, or electrical generation in new or existing residential or commercial buildings are excluded from business activities tax. Expires 1/1/85.	State Department of Treasury State Tax Commission State Capitol Building Lansing, MI 48922 (517) 373-2910
	PA 133, 1976	Tangible property used for solar, wind, or water energy devices is excluded from excise tax if it is used to heat, cool, or electrify a new or existing commercial or residential building. Expires 1/1/85.	State Department of Treasury State Tax Commission State Capitol Building Lansing, MI 48922 (517) 373-2910
	PA 135, 1975	This law exempts solar, wind, or water energy conversion devices from real and	Local Government Services Treasury Building

		personal property tax. An application must be filed with local tax assessor, who will submit it to the state tax commission for certification. Authority to exempt expires 7/1/85, but exemptions made by that time stay in force.	Lansing, MI 48922 (517) 373-3232
Michigan	PA 605, 1979	Income tax credit may be claimed for a residential solar, wind, or water energy device that is used for heating, cooling, or electricity. This includes devices designed to use the difference between water temperatures in a body of water. Energy conservation measures installed in connection with such devices are also eligible; these include insulation and water-flow reduction devices. Swimming pool heaters are eligible only if 25% or more of their heating capacity is used for residential purposes. The credit may be carried from year to year until it is expended. The law instructs the Department of Commerce to establish system eligibility standards within 180 days of the law's passage. To be eligible, expenditures must be made by 12/31/83.	State Department of Treasury State Tax Commission State Capitol Building Lansing, MI 48922 (517) 373-2910

State	Law	Description	Contact
		The rate of credit changes annually. For 1979, the rate for single-family dwellings is 25% of the first \$2000 spent, plus 15% of the next \$8000 spent. In 1979, the rate for other buildings is 25% of the first \$2000, plus 15% of the next \$13,000.	
Minnesota	Ch. 786 Laws of 1978	The market value of agriculturally derived methane gas, solar, or wind systems used for heating, cooling, or electricity in a building or structure is excluded from property tax. The installations must be done prior to 1/1/84.	Local assessor or board of assessors
Montana	Ch. 548 Laws of 1975 as amended by Ch. 574 Laws of 1977	Energy systems using non-fossil fuel energy (such as solar, wind, decomposition of organic wastes) installed in a taxpayer's dwelling before 12/31/82 are eligible for a tax credit of 10% for the first \$1000 and 5% of the next \$3000. If a federal tax credit is also claimed, the state credit is reduced to 5% of the first \$1000 and 2½% of the next \$3000.	State Department of Revenue Income Tax Section Mitchell Building Helena, MT 59601 (406) 449-2837

	Ch. 576 Laws of 1977	This law provides individual or corporate deductions for energy conservation improvements, including storm windows and insulation. It applies to all types of buildings at the following rates:	State Department of Revenue Income Tax Section Mitchell Building Helena, MT 59601 (406) 449-2837
Nevada	Ch. 345 Laws of 1977	This law establishes a property tax allowance on solar, wind, geothermal, water-powered, or solid waste energy systems in residential buildings. The property tax allowance equals the difference in tax on the property with the energy system and the tax on the property without the energy system. The allowance may not exceed the tax accrued or $2000, whichever is less. Claims are to be filed with the county assessor.	Local county assessor

Rates for Ch. 576 Laws of 1977:

Residential buildings	Non-residential buildings
100% of 1st $1000	100% of 1st $2000
50% of 2nd $1000	50% of 2nd $2000
20% of 3rd $1000	20% of 3rd $2000
10% of 4th $1000	10% of 4th $2000

State	Law	Description	Contact
New Hampshire	Ch. 391 Laws of 1975, as amended by Ch. 5202 Laws of 1977	Cities and towns are enabled to grant a property tax exemption to property owners with solar heating, cooling, or hot water systems and will decide the amount of the exemption and the manner of determination. An application for the exemption must be filed with the local assessor.	Local assessor or board of assessors
New Jersey	Ch. 256 Laws of 1977	Solar heating and cooling systems, including sea thermal gradients and wind-powered systems, are exempt from property tax. Systems must be certified under the State Uniform Construction Act on forms designated by the Division of Taxation. Systems must meet standards established by the State Energy Office. Expires 12/31/82.	Local assessor or board of assessors
	Ch. 465 Laws of 1977	Solar energy devices designed to provide heating, cooling, electrical or mechanical power are exempt from sales tax. These systems must meet the standards established by the state Department of Energy.	State Division of Taxation Tax Counselors P.O. Box 999 Trenton, NJ 08646 (609) 292-6400

New Mexico	Ch. 12 Laws of 1975, as amended by Ch. 170 Laws of 1978	This law provides for an income tax credit of 25% of the cost of a solar energy system or a maximum of $1000. It is available for solar energy systems which heat and/or cool the taxpayer's residence and for swimming pool heating systems. The credit is not available if federal credit is claimed. The criteria of the Solar Heating and Cooling Act of 1974 (42 USC 5506) must be met. Credit in excess of the taxes due will be refunded.	State Department of Taxation & Revenue, Income Tax Division P.O. Box 630 Sante Fe, NM 87503 (505) 827-3221
	Ch. 114 Laws of 1977	Individuals may claim an income tax credit for a solar energy system used in an irrigation pumping system. The system design must be approved by the energy resources board prior to installation, and it must result in a 75% reduction in the use of fossil fuel. This law is not applicable if federal credit were claimed or if credit were claimed for this equipment under other provisions of the state law. Credit in excess of the taxes due will be refunded.	State Department of Taxation & Revenue, Income Tax Division P.O. Box 630 Santa Fe, NM 87503 (505) 827-3221

State	Law	Description	Contact
New York	Ch. 322 Laws of 1977	This law provides for property tax reductions for owners of solar or wind energy systems. The reduction of assessment is equal to the difference between assessment of the property with the energy system and assessment of the property without the system. The system must conform to guidelines of the state energy office and must be installed before 7/1/88. The exemption is good for 15 years after it is granted.	Local assessor or board of assessors
North Carolina	Ch. 792 Laws of 1977	This law provides for a corporate and individual income tax credit of 25% of the cost of a solar heating, cooling, or hot water system. There is a maximum credit of $1000 per unit or building. Although this credit may be taken only once, the amount of credit may be spread over 3 years. The system may be in any type of building, and it must meet the criteria of the Solar Heating and Cooling Demonstration Act of 1974.	State Department of Revenue Income Tax Division P.O. Box 25000 Raleigh, NC 27540 (919) 733-3991

	Ch. 965 Laws of 1977	Buildings with solar heating or cooling systems shall be assessed as though they had a conventional system. Expires 12/1/85.	Local assessor or board of assessors
North Dakota	Ch. 537 Laws of 1977	This law provides for an income tax credit for solar or wind energy devices. The credit is 5% per year for 2 years. The system must provide heating, cooling, mechanical, or electrical power.	State Tax Commission Income Tax Division Capitol Building Bismarck, ND 58505 (701) 224-3450
	Ch. 508 Laws of 1975	Solar heating or cooling systems in any building are exempt from property tax for 5 years after installation.	Local assessor or board of assessors
Oklahoma	Ch. 209 Laws of 1977	Individuals may claim an income tax credit for solar energy devices used to heat, cool, or furnish electrical or mechanical power at principal residence. The credit is equal to 25% of the cost of the system or a maximum of $2000. Although this credit may be taken only once, the amount of credit may be spread over 3 years. Expires on 1/1/88.	State Tax Commission Income Tax Division 2501 Lincoln Boulevard Oklahoma City, OK 73194 (405) 521-3125

State	Law	Description	Contact
Oregon	Ch. 196 Laws of 1977	This law provides for income tax credit for the installation of solar, wind, or geothermal energy systems in a dwelling for purposes of heating, cooling, hot water, or electrical power. The credit equals 25% of the cost of the system. The maximum credit is $1000. Although this credit may be taken only once, the amount of credit may be spread over 3 years. The systems must meet the criteria of the Department of Energy. The credit expires on 1/1/85.	State Department of Revenue State Office Building Salem, OR 97310 (503) 378-3366
		This law also provides for a property tax exemption for property equipped with solar energy systems. The exemption equals the value of the property with the solar energy system minus the value of the property without the system. This credit expires 1/1/98.	Local assessor or board of assessors
Rhode Island	Ch. 202 Laws of 1977	Solar heating or cooling systems in residential or non-residential buildings shall be assessed at no more than the value of a conventional system necessary to serve the building. Law expires 4/1/97.	Local assessor or board of assessors

South Dakota	Ch. 74 Laws of 1978	This law provides property tax assessment credit for renewable resource energy systems (sun, wind, geothermal, and biomass). For residential property, the amount of the credit equals the assessed value of the property with the system, minus the assessed value of the property without the system, but not less than the actual installation cost of the system. The credit for systems in commercial buildings is equal to 50% of the cost of installation. For residential buildings, full credit is given for 5 years. For the next 3 years, the credit is 75%, 50%, and 25% of the full credit. For commercial buildings, full credit is given for 3 years, and for the next 3 years credit is 75%, 50%, and 25% of the full credit. Taxpayers must apply to the county auditor.	Local county assessor
Tennessee	Ch. 837 Laws of 1978	Solar or wind energy systems for heating, cooling, or electrical power shall be exempt from property taxation. Law expires 1/1/88.	Local assessor or board of assessors

State	Law	Description	Contact
Texas	Article VIII, Sec. 2(a) of Texas Constitution 1978	The legislature is allowed to exempt solar- or wind-powered energy devices from property tax.	
	Ch. 719 Laws of 1975	Solar energy systems used for heating, cooling, or electrical power are exempt from sales tax. Corporations may deduct from taxable capital the amortized cost of a solar energy device over a period of 60 months or more.	Comptroller of Public Accounts Capitol Station, Drawer SS Austin, TX 78775 (512) 475-2206
Texas	Ch. 584 Laws of 1977	This law provides a franchise tax exemption for corporations exclusively engaged in manufacturing, selling, or installing solar energy devices for heating, cooling, or electrical power.	Comptroller of Public Accounts Capitol Station, Drawer SS Austin, TX 78775 (512) 475-2206
Vermont	Act 226 Laws of 1976	Towns may enact a property tax exemption for alternate energy systems. Systems exempted are grist mills, windmills, solar energy systems, and devices to convert organic matter to methane. All components are exempt, including land on which the facility is situated, up to ½-acre.	Local assessor or board of assessors

	Act 210 Laws of 1978	Wood-fired central heating and solar or wind systems for heating, cooling, or electrical power are eligible for income tax credit if they are installed in the taxpayer's dwelling before 7/1/83. The credit is equal to 25% of the cost of the system or $1000, whichever is less. Businesses may deduct 25% of the cost of the system or $3000, whichever is less.	State Tax Department Income Tax Division State Street Montpelier, VT 05602 (802) 828-2517
Virginia	Ch. 561 Laws of 1977	Any county, city, or town may exempt solar energy equipment used for heating, cooling, or other applications from property tax. The State Board of Housing must certify the system. The exemption is good for not less than 5 years.	Local tax governing body
Washington	Ch. 364 Laws of 1977	Solar water and space heating or solar power systems are exempt from property taxation. Claims must be filed with the county assessor. The exemption is valid for 7 years. Claims must be filed by 12/31/81.	Local assessor or board of assessors

State	Law	Description	Contact
Wisconsin	Ch. 313 Laws of 1977	Alternative energy systems installed by businesses may be used as a tax deduction in the year paid for, may be depreciated, or may be amortized over 5 years. This law covers solar, waste conversion, and wind systems but does not include solid fuel consuming devices used for residential purposes. The system must be certified by the Department of Industry, Labor and Human Relations. Individuals may claim income tax credit on systems costing at least \$500. The credit may not exceed \$10,000. The following tax credit applies to systems installed on buildings that existed before 4/20/77: If the system were installed in: — The tax credit equals this % of the system cost: 1979, 1980 — 24% 1981, 1982 — 18% 1983, 1984 — 12% For buildings constructed after 4/20/77 the following credit rate applies:	State Department of Revenue Income Tax Division 3648 University Avenue Madison, WI 53705 (608) 266-2772

If the system were installed in:	The tax credit equals this % of the system cost:
1979, 1980	16%
1981, 1982	12%
1983, 1984	8%

Any tax credit in excess of taxes due may be refunded. Expenses must be incurred before 12/31/84.

B. Grants and Loans

State	Law	Description	Contact
California	Ch. 1 and Ch. 7 Laws of 1978	This law creates the Solar Energy Demonstration Loan Program that provides $2000 interest-free loans for solar space heating and domestic hot water systems in areas where a state of emergency has been declared.	Department of Housing and Community Development Division of Research and Policy Development 921 10th Street, Fifth Floor Sacramento, CA 95814 (916) 445-4728
	Ch. 1243 Laws of 1978	The maximum loan available to veterans for home mortgages is increased if the home is equipped with a solar energy system.	Department of Veterans' Affairs 1227 "O" Street, P.O. Box 1559 Sacramento, CA 95807 (916) 445-2347 or any local district office of the Department of Veterans' Affairs

State	Law	Description	Contact
Illinois	PA 80-430, 1977	This law establishes a $5 million solar energy demonstration program.	Division of Energy Illinois Dept. of Business & Economic Development 222 South College Springfield, IL 62706 (217) 782-7500
Iowa	Ch. 1086 Laws of 1978	This law establishes a loan and grant fund for property improvements and mortgages for low-income families. Solar energy systems qualify as improvements.	Executive Director, Iowa Housing Finance Authority 218 Liberty Building Des Moines, IA 50319 (515) 281-4058
Maine	Ch. 685 Laws of 1978	Maine provides $16,000 for a solar hot water demonstration program; grants are $400 each.	Office of Energy Resources 55 Capitol Street Augusta, ME 04330 (207) 289-2196
Massachusetts	Ch. 28 Laws of 1977, Ch. 260 Laws of 1977, Ch. 73 Laws of 1978	Banks and credit unions are authorized to make loans with extended maturation periods and increased maximum amounts for home improvements, including solar energy systems. Banks may lend up to $15,000, and credit unions may lend up to $12,000.	Local bank or credit union

Montana	Ch. 548 Laws of 1975	This law allows a public utility to lend money for the installation of solar energy systems in dwellings at an interest rate not to exceed 7% per year. The utility will then receive a license tax credit for the difference between 7% and the prevailing rate of interest.	Public Service Commission 1227 11th Avenue Helena, MT 59601 (406) 449-3008
Oregon	Ch. 732 Laws of 1977	This law creates a loan fund for alternate energy projects and authorizes the Director of the Department of Energy to sell bonds to finance the loan fund.	Oregon Dept. of Energy Rm. 111 Labor & Industries Bldg. Salem, OR 97310 (503) 378-4128
	Ch. 315 Laws of 1977	To finance a domestic solar energy system, veterans can obtain a loan in excess of the maximum allowed under the War Veterans Fund. The system must provide at least 10% of the home's energy requirements and must meet performance criteria established by the State Department of Energy.	State Dept. of Veterans' Affairs 3000 Market St. Plaza, Suite 522 Salem, OR 97310 (503) 378-6438
Tennessee	Ch. 884 Laws of 1978	Provides loans to low- and moderate-income persons to make energy conserving improvements, including the installation of solar hot water systems.	Tennessee Housing Development Authority Hamilton Bank Building Nashville, TN 37219 (615) 741-3023

C. Land Use Provisions

State	Law	Description	Contact
California	Ch. 1366 Laws of 1978	Anyone who owns, occupies, or controls real estate is prohibited from allowing a tree or shrub to cast a shadow on a solar collector between 9:30 a.m. and 2:30 p.m. Trees casting a shadow before the installation of a collector are excluded.	
	Ch. 1154 Laws of 1978	This law declares that any restriction on real property purporting to prohibit the installation and use of a solar energy system is void and unenforceable. It recognizes solar easements and prescribes their contents. City and county governments may not prohibit or restrict solar energy systems except to ensure the public health. The law requires that new subdivision maps be designed to accommodate passive solar energy systems to the maximum extent possible. It permits city and county governments to require the dedication of solar easements before approving the map.	Local planning, zoning body

Colorado	Ch. 326 Laws of 1975	Solar easements are recognized and their contents are prescribed.	Local clerk or recorder
Connecticut	PA 314 Laws of 1978	The zoning commission of each city, town, or borough is authorized to regulate development to encourage energy efficiency and the use of renewable forms of energy, including solar.	Local planning/zoning body
Florida	Ch. 309 Laws of 1978	Solar easements are recognized and subject to the same requirements as other easements; the contents are prescribed.	Local clerk or recorder
Georgia	Act 1446 Laws of 1978	Solar easements are recognized and subject to the same requirements as other easements; the contents are prescribed.	Local clerk or recorder
Idaho	Ch. 294 Laws of 1978	Solar easements are recognized and are made subject to the same requirements as other easements; the contents are prescribed.	Local clerk or recorder
Kansas	Ch. 227 Laws of 1977	Solar easements are recognized and are subject to the same requirements as other easements; the contents are prescribed.	Local clerk or recorder
Maryland	Ch. 934 Laws of 1977	Solar easements are recognized as a lawful restriction on land.	Local clerk or recorder

State	Law	Description	Contact
Minnesota	Ch. 786 Laws of 1978	Zoning ordinances may provide for the protection of solar access for solar energy energy systems. Solar easements are recognized and the contents are prescribed; they are enforceable in civil actions. Depreciation resulting from easements (but not any appreciation) shall be included in revaluation for property tax.	Local planning/zoning body (For zoning) Local clerk or recorder (For easements)
New Jersey	Ch. 152 Laws of 1978	Solar easements are recognized and subject to the same requirements as other easements; the contents are prescribed.	Local clerk or recorder
New Mexico	Ch. 169 Laws of 1977	The right to use solar energy is a property right of landowners; disputes regarding access will be settled by rule of prior appropriation.	Local municipal or county zoning authority
North Dakota	Ch. 425 Laws of 1977	Solar easements are recognized and subject to the same requirements as other easements; the contents are prescribed.	Local clerk or recorder
Virginia	Ch. 323 Laws of 1978	This law subjects solar easements to the same legal requirements as other easements and mandates contents of the agreement.	Local clerk or recorder

D. Standards and Regulation of Construction

State	Law	Description	Contact
California	Ch. 670 Laws of 1976	Any city or county may require that new buildings subject to the State Housing Law be constructed in a manner that permits the installation of solar heating.	Local building inspector
	Ch. 1081 Laws of 1977	The State Energy Resources, Conservation and Development Commission is required to adopt regulations and standards governing solar energy equipment.	Energy Resources Conservation and Development Commission 1111 Howe Avenue Sacramento, CA 95825 (916) 322-3690
Connecticut	PA 76-409, 1976	The Commissioner of Planning and Energy Policy is required to establish standards for solar energy systems.	Connecticut Energy Div. — OPM 20 Grand Street Hartford, CT 06115 (203) 566-5765
Florida	Ch. 76-246 Laws of 1976	The Florida Solar Energy Center is required to establish standards for solar energy systems.	Florida Solar Energy Center 300 State Road 401 Cape Canaveral, FL 32920 (305) 783-0300
	Ch. 74-361 Laws of 1974	The law stipulates that no single-family dwelling shall be constructed unless it is designed to facilitate future installation of a solar hot water system.	Local building inspector

State	Law	Description	Contact
	Ch. 78-309 Laws of 1978	All solar energy systems manufactured or sold in Florida must meet the standards established by the Florida Solar Energy Center.	Bureau of Codes and Standards 2571 Executive Center Circle Tallahassee, FL 32301 (904) 488-3581
Illinois	PA 80-430, 1977	The Division of Energy of the Department of Business and Economic Development is required to establish guidelines and regulations for solar energy systems.	Illinois Institute of Natural Resources 325 West Adams, Room 300 Springfield, IL 62706 (217) 785-2800
Louisiana	Act 542, 1978	The law requires the Department of Natural Resources and Development to adopt regulations and standards governing solar energy devices.	Dept. of Natural Resources and Development P.O. Box 44396 Baton Rouge, LA 70804 (504) 342-4500
Michigan	PA 605, 1979	The Department of Commerce is required to formulate standards for solar energy systems.	Energy Extension Service Michigan Energy Administration Dept. of Commerce P.O. Box 30228 Lansing, MI 48909 (517) 373-6430

Minnesota	Ch. 333 Laws of 1976	The Building Code Division of the Department of Administration is required to promulgate performance standards for solar energy systems. (SBC 6101-6108)	Minnesota Department of Administration, Bldg. Code Div. 408 Metro Square Building St. Paul, MN 55101 (612) 296-4639
Nevada	Ch. 513 Laws of 1977	The law stipulates that the Public Works Board shall establish standards for new buildings. The standards shall include provisions for the design and construction of solar energy systems.	State Public Works Board 505 East King Street, Room 400 Carson, City, NV 89710 (702) 885-4870
New Jersey	Ch. 256 Laws of 1977	The Division of Energy Planning and Conservation of the State Energy Office is required to adopt standards for solar energy systems.	Department of Energy 101 Commerce Street Newark, NJ 07102 (201) 648-3290
New Mexico	Ch. 347 Laws of 1977	This law directs the New Mexico Solar Energy Research and Development Institute to develop performance standards for solar energy equipment.	New Mexico Solar Energy Research and Development Institute, Box 3 SOL New Mexico State University Las Cruces, NM 88003 (505) 646-1846

State	Law	Description	Contact
New York	Ch. 322 Laws of 1977	This law requires that the Commissioner of the State Energy Office promulgate guidelines and definitions for solar energy systems.	New York State Energy Office Agency Bldg. 2 Empire State Plaza Albany, NY 12223 (518) 474-8181
Oregon	Ch. 196 Laws of 1977	The Department of Energy is required to adopt rules prescribing performance criteria for solar energy systems.	Oregon Dept. of Energy Rm. 111, Labor & Industries Bldg. Salem, OR 97310 (503) 378-4128

3. State Departments of Energy

Alaska

Department of Commerce & Economic Development
Division of Energy & Power Development
338 Denali Street
Anchorage, AK 99501
(907) 272-0527

Activities/Services: Conducts policy planning, alternative energy studies, wind demonstration projects, surveys and workshops; maintains solar advisory group.
Distributes consumer information.

Alabama

Alabama Solar Center
Environmental & Energy Center
University of Alabama/Huntsville
PO Box 1247, Annex D
Huntsville, AL 35807
(205) 895-6257

Activities/Services: Tests collectors and systems; conducts workshops and exhibits; maintains information service and library; answers solar questions.
Distributes own test reports and studies, state solar directory, and other publications.

Arizona

Solar Information
Arizona Solar Energy Research Commission
Room 502 — Capitol Tower
1700 W. Washington
Phoenix, AZ 85007
(602) 271-3682 Solar Information
*(800) 352-5499 State Energy Hotline
(602) 271-3303 State Energy Hotline (Phoenix only)

*Toll-free number.

Activities/Services: Conducts policy planning, feasibility studies, state demonstration projects, conferences and workshops (for general public and solar or design professionals). Distributes state solar directory, consumer guide, and federal solar publications.

Arkansas

Energy Information
Arkansas Energy Office
960 Plaza West
Little Rock, AR 72205
(501) 371-1370 Solar Information
*(800) 482-1122 State Energy Hotline

Activities/Services: Answers solar and general energy-related questions; distributes fact sheets and bibliographies. Conducts workshops for general public and civic organizations.

California

Energy Information
California Energy Commission
1111 Howe Avenue
Sacramento, CA 95825
*(800) 852-7516 State Solar Hotline

Activities/Services: Answers solar and other alternative energy questions, state solar tax credit inquiries, consumer complaints. Tests and certifies collectors; conducts policy planning; develops code manuals for building and planning officials. Distributes list of California solar professionals, information on state solar tax credit, and solar publications.

Colorado

Energy Information
Office of Energy Conservation
1410 Grant Street
Suite B-104
Denver, CO 80203
(303) 839-2507

Activities/Services: Conducts policy planning; develops educational programs (vocational and other levels), conducts workshops. Distributes solar information and publications of statewide interest. Offers information on eight regional energy extension centers located throughout the state.

Connecticut

Solar
Office of Policy & Management
Energy Division
20 Grand Street
Hartford, CT 06115
(203) 566-5803 Solar Information
*(800) 842-1648 State Energy Extension Service

Activities/Services: Conducts policy planning, feasibility studies; offers information on state solar tax incentives, courses, and consumer protection; monitors local HUD Residential Solar Demonstration projects. Offers *Connecticut Energy Outlook*, an assessment of the state's energy supply and demand (with recommendations for legislative action), in summary form.

*Toll-free number.

Delaware

Governor's Energy Office
PO Box 1401
Townsend Building
Dover, DE 19901
*(800) 282-8616 State Energy Hotline

Activities/Services: Conducts policy planning, feasibility studies, certifies systems for state solar tax credit; offers training programs for installers and building inspectors. Monitors local HUD Residential Solar Demonstration projects. Distributes installation guidelines and selected federal solar publications. Develops educational material for public schools.

Florida

Florida Solar Energy Center
300 State Road 401
Cape Canaveral, FL 32920
(305) 783-0300

Activities/Services: Conducts technical and market research, tests collectors, conducts training programs for installers. Monitors local HUD Residential Solar Demonstration projects. Demonstrates consumer and technical publications; answers solar questions.

Georgia

Solar Energy Coordinator
Georgia Office of Energy Resources
Room 615
270 Washington Street, SW
Atlanta, GA 30334
(404) 656-5176

Activities/Services: Conducts research and policy planning, answers solar and other energy-related inquiries.

Hawaii

State Energy Office
Department of Planning & Economic Development
PO Box 2359
Honolulu, HI 96804
(808) 548-4195
(808) 548-4150
(808) 548-4080
State Energy Hotline
*Dial 0 — ask for 8016 (toll-free within Hawaii)

Activities/Services: Conducts policy planning, workshops, conferences, and seminars; assesses new alternative energy technologies.
Develops and distributes publications on solar energy and conservation.

Idaho

Energy Information
Office of Energy
State House
Boise, ID 83720
(208) 384-3258

Activities/Services: Answers solar and other energy-related questions.
A monthly general energy-related newsletter is available upon request.

*Toll-free number.

Illinois

Energy Information Center
Institute of Natural Resources
222 S. College
Springfield, IL 62706
(217) 782-1926

Activities/Services: Conducts policy planning, feasibility studies, develops vocational education materials, and sponsors design and installer workshops. Answers solar and other energy-related questions. Ask for a copy of the center's publication list.

Indiana

Solar
Indiana Department of Commerce
Energy Group
7th Floor — Consolidated Building
115 N. Pennsylvania Street
Indianapolis, IN 46204
(317) 633-6753

Activities/Services: Conducts policy research, answers solar and general energy-related questions.

Iowa

Solar
Energy Policy Council
215 E. 7th Street
Des Moines, IA 50319
(515) 281-4420

Activities/Services: Conducts policy planning, coordinates workshops, administers state solar research and development projects, develops energy education materials.
Distributes state solar directory, annual summary of state and federal solar programs, survey of solar legislation, and selected federal publications.

Kansas

Kansas Energy Office
Room 241
503 Kansas Avenue
Topeka, KS 66603
(913) 296-2496

Activities/Services: Answers general energy-related questions (including solar). Assists residents in filing applications for federal energy-related grant programs. Distributes state solar directory and federal solar publications.

Kentucky

Bureau of Energy Management
Kentucky Department of Energy
Capitol Tower
Frankfort, KY 40601
*(800) 372-2978
State Energy Hotline

Activities/Services: Conducts policy planning, seminars and workshops. Answers solar and other energy-related questions. Publishes do-it-yourself solar information and distributes federal solar publications.

Louisiana

Research & Development Division
Department of Natural Resources
PO Box 44156
Baton Rouge, LA 44156

*Toll-free number.

Activities/Services: Conducts policy planning, administers state solar cooling demonstration projects.
Distributes solar information packages to public libraries across the state.

Maine

Maine Office of Energy Resources
55 Capitol Street
Augusta, ME 04330
(207) 289-2195

Activities/Services: Conducts policy planning; administers state solar hot water grant program. Distributes federal solar publications.

Maryland

Solar
Maryland Energy Office
Room 1302
301 W. Preston Street
Baltimore, MD 20201
(301) 383-6810 Solar Information
*(800) 492-5903
Home Energy Savers Program

Activities/Services: Assists residents filing for federal energy-related grants; offers information on state solar legislation; monitors local HUD Residential Solar Demonstration projects.
Distributes lists of solar professionals. General energy newsletter available.

Massachusetts

Solar Action Office
Room 1413
One Ashburton Place
Boston, MA 02108
(617) 727-7297 Solar Information
*(800) 922-8265
State Energy Hotline

Activities/Services: Conducts policy planning, feasibility studies, and workshops. Develops guidelines for building codes and property tax assessments; monitors local HUD Residential Solar Demonstration projects. Handles consumer problems (in cooperation with state attorney general's office), answers technical and nontechnical questions, distributes state and federal solar publications.

Michigan

Solar Energy
Division of Business & Renewable Resources
Energy Administration
Department of Commerce
4th Floor — Law Building
Lansing, MI 48913
(517) 373-6430 Solar Information
*(800) 292-4704
State Energy Extension Service

Activities/Services: Conducts policy planning, feasibility studies, conferences and workshops. *Alternate Energy Guidebook* (includes photo directory of solar buildings in Michigan).

Minnesota

Minnesota Energy Agency
980 American Center Building
150 E. Kellogg

*Toll-free number.

St. Paul, MN 55101
(612) 292-8382 Solar Information
*(800) 652-9028
State Energy Hotline

Activities/Services: Conducts policy planning; distributes state solar directory and list of state solar buildings.

Mississippi

Solar Energy
Mississippi Fuel & Energy Management Commission
1307 Woolfolk Building
Jackson, MS 39202
(601) 354-7406

Activities/Services: Monitors solar lumber-drying project; studies "solar barriers," and co-sponsors "solar barriers" workshops open to the public. Distributes federal solar publications and other energy information.

Missouri

Solar Information
Missouri Energy Program
PO Box 1309
Jefferson City, MO 55102
*(800) 392-8269
State Energy Hotline

Activities/Services: Conducts feasibility studies; distributes basic solar primer and blueprints for active and passive systems.

Montana

Energy Division
Department of Natural Resources & Conservation
Capitol Station
Helena, MT 59601
(406) 449-3940

Activities/Services: Conducts policy planning, feasibility studies, and seminars; administers state solar grant program (includes solar heating, wind energy, and biomass).
Solar handbook.

Nebraska

Nebraska State Solar Office
W-191, Nebraska Hall
University of Nebraska
Lincoln, NB 68588
(402) 472-3414

Activities/Services: Conducts workshops; distributes state solar directory, federal solar publications, and a general energy newsletter.

Nevada

Solar
Department of Energy
1050 E. Williams
Suite 405
Capitol Complex
Carson City, NV 89710
(702) 885-5157

Activities/Services: Conducts policy planning; coordinates research programs with Desert Research Institute; conducts workshops.
Distributes federal solar publications, state solar consumer guide.

*Toll-free number.

New Hampshire

Solar
Governor's Council on Energy
26 Pleasant
Concord, NH 03301
(603) 271-2711 Solar Information
*(800) 562-1115
State Energy Hotline

Activities/Services: Monitors local HUD Residential Solar Demonstration projects; conducts feasibility studies.
Distributes state solar directory, local solar hot water information packet, and federal solar publications.

New Jersey

Office of Alternative Technology
State Energy Office
101 Commerce Street
Newark, NJ 07102
(201) 648-6293 Solar Information
*(800) 492-4242
State Energy Hotline

Activities/Services: Studies solar and solid-waste recovery energy sources; conducts policy planning, feasibility studies, seminars and workshops for building inspectors and installers. Monitors local HUD Residential Solar Demonstration projects. State solar directory, lists of solar buildings, and a general consumer primer.

New Mexico

Solar Energy &
Minerals Department
PO Box 2770
Santa Fe, NM 87501
(505) 827-2472 Solar Information
(505) 827-2386
State Energy Extension Service
*(800) 432-6782/3
State Energy Hotline

Activities/Services: Conducts policy planning, feasibility studies, and workshops; tests collectors; administers state solar research and development program. Administers research-oriented Solar Energy Institute operated by New Mexico State University in cooperation with the Energy & Minerals Department. Energy Extension Service answers technical and nontechnical questions, offers information on state solar tax laws, and distributes introductory primers on active, passive, hybrid solar hot water and space heating systems. Contact them for information on short courses for building attached greenhouses.

New York

New York State Energy Office
Solar Program
Agency Building #2
Empire State Plaza
Albany, NY 12223
(518) 474-7016
New York State Energy Hotline

Activities/Services: Conducts policy planning; sponsors conferences and workshops; maintains inventory of solar manufacturers, dealers, and installers. Assists

*Toll-free number.

local governments in promotion of solar energy; maintains Solar Energy Applications Center (SEAC); monitors local HUD Residential Solar Demonstration projects. Provides consumer information; distributes solar energy and conservation publications.

North Carolina

Information Section
Energy Division
North Carolina Department
of Commerce
PO Box 25249
Raleigh, NC 27611
*(800) 662-7131
State Energy Hotline

Activities/Services: Conducts policy planning; conducts research jointly with the University of North Carolina to formulate solar training materials for high school vocational courses; conducts solar energy workshops for low income residents. Distributes state solar directory and federal solar publications.

North Dakota

Energy Information
Office of Energy Management
& Conservation
1533 N. 12th Street
Bismark, ND 58501
(701) 224-2250

Activities/Services: Conducts workshops; answers questions on energy (including solar).

Ohio

Solar Information
Ohio Department of Energy
34th Floor — State Office Tower
30 E. Broad Street
Columbus, OH 43215
(614) 466-8102 Solar Information
*(800) 282-9234
State Energy Hotline

Activities/Services: Conducts policy planning, seminars and feasibility studies of solar industrial process heat in cooperation with Ohio State University. Publishes *Energy Saving Home Improvement Guide*, available through toll-free number.

Oklahoma

Energy Information
Oklahoma Department of Energy
4400 N. Lincoln Boulevard
Suite 251
Oklahoma City, OK 73105
(405) 521-3941

Activities/Services: Conducts policy planning, feasibility studies. Answers solar and general energy-related questions.

Oregon

Energy Information
Oregon Department of Energy
Room 111
Labor & Industries Building
Salem, OR 97310
*(800) 452-7813 "Access 800"
State Government Hotline

*Toll-free number

Activities/Sercices: Conducts policy planning, feasibility studies, and workshops; certifies active and passive systems for state solar tax credit; answers questions referred from "Access 800."

Pennsylvania

Governor's Energy Council
500 N. 3rd Street
9th Floor
Harrisburg, PA 17101
*(800) 882-8400
State Energy Extension Service

Activities/Services: Monitors local HUD Residential Solar Demonstration projects; conducts market research; inventories state energy resources; studies feasibility of solar retrofits for public buildings; conducts workshops; develops solar consumer publications.

Rhode Island

Governor's Energy Office
80 Dean Street
Providence, RI 02903
(401) 277-3370 Solar Information
(401) 277-3773/4
State Energy Hotline

Activities/Services: Conducts policy planning and coordination, solar economic studies, energy conservation training, and workshops. Maintains energy information center; answers technical and nontechnical questions. Publishes *Southern New England Solar Directory*, list of solar buildings in Rhode Island, solar and alternative energies handbook, and other publications.

South Carolina

Solar Energy
South Carolina Energy Management Office
Edgar A. Brown Building
Room 306
1205 Pendelton Street
Columbia, SC 29201
(803) 758-2050 Solar Information
*(800) 922-1600
State Energy Hotline

Activities/Services: Develops energy conservation courses for vocational schools; conducts seminars and policy planning.

South Dakota

State Solar Office
Suite 102
Capitol Lake Plaza Building
Pierre, SD 57501
*(800) 592-1865 "Tie Line"
State Energy Hotline

Activities/Services: Conducts feasibility studies; gathers climatic data. Answers solar and general energy-related questions.

Tennessee

Solar
Tennessee Energy Authority
250 Capitol Building
Nashville, TN 37219
(615) 741-2994 Solar Information
*(800) 642-3281
State Energy Hotline

*Toll-free number

Activities/Services: Studies existing state solar buildings; conducts feasibility studies; organizes workshops on passive solar design for builders, developers, and engineers. Offers information on the TVA Solar Hot Water demonstration program (Memphis area only).

Texas

Librarian
Texas Energy Advisory Council
7703 N. Lamar Street
Austin, TX 78752
(512) 475-5588 Solar Information
(713) 845-8025
State Energy Extension Service

Activities/Services: Alternative energy research and development; conducts policy planning, feasibility studies. Distributes research program reports, quarterly newsletter on state Energy Advisory Council activities, and other publications. Ask for their publications list.

Utah

Engineering Experiment Station
105 Applied Research Building
University of Utah
Salt Lake City, UT 84112
*(800) 662-3633
State Energy Hotline

Activities/Services: Conducts policy planning; answers solar and general energy-related questions.

Vermont

Solar
State Energy Office
Pavilion Office Building
Montpelier, VT 05602
*(800) 642-3281
State Energy Hotline

Activities/Services: Advises state legislature; conducts policy research; studies feasibility of solar energy retrofits for public buildings; monitors local HUD Residential Solar Demonstration projects. Answers questions and offers solar advice to consumers.

Virginia

Virginia Energy Information
& Services Center
310 Turner Road
Richmond, VA 23235
*(800) 552-3831
State Energy Hotline

Activities/Services: Maintains solar energy library; answers technical and nontechnical questions. Distributes fact sheets, bibliographies, and state solar directory.

Washington

Solar
State Energy Office
400 E. Union Street
1st Floor
Olympia, WA 98504
(206) 754-1350 Solar Information
(206) 344-3440
State Energy Extension Service

*Toll-free number.

Activities/Services: Conducts policy planning, feasibility studies, regional conferences and workshops.
Fact sheets, bibliographies, and do-it-yourself plans.
Schedules the *Energizer,* a traveling exhibit on energy conservation.

West Virginia

Solar Energy
Fuel & Energy Office
1262½ Greenbrier Street
Charleston, WV 25311
*(800) 642-9012/3
State Energy Hotline

Activities/Services: Offers state directory of solar architects and engineers; distributes federal solar publications.

Wisconsin

Office of Planning & Energy
State Office Building
1 W. Wilson, Room 201
PO Box 511
Madison, WI 53701
(608) 266-3382
Write for information.

Activities/Services: Conducts policy planning, data surveys, state solar installations survey, and solar conferences; maintains Solar Resource Advisory Panel; distributes solar energy packet (developed jointly with State Energy Extension Service). Publishes *Wisconsin Solar News* bimonthly, available upon written request.

Wyoming

Energy Information
PO Box 3965
University Station
Laramie, WY 82070
*(800) 442-6783
State Energy Hotline

Activities/Services: Conducts policy planning, feasibility studies, workshops, develops educational materials; administers state appropriate technology seed grant program. Maintains energy information center in conjunction with University of Wyoming.

*Toll-free number

4. Energy Extension Services by State

At present there are ten states where pilot Energy Extension Services are being developed. These extension services will answer specific questions, as well as inquiries on where to find more information. Modeled after the agricultural extension services, this pilot program is being funded by the Department of Energy.

Alabama

Energy Extension Service
Auburn University
Auburn, AL 36830

Connecticut

Energy Extension Service
Office of Policy and Management
20 Grand Street
Hartford, CT 06115

Michigan

Energy Extension Service
209 Hollister Building
Lansing, MI 48933

New Mexico

Energy Extension Service
PO Box 00, 440 Cerrillos Road
Santa Fe, NM 87501

Pennsylvania

Governor's Energy Council
State Street Building, 12th Floor
Harrisburg, PA 17101

Tennessee

Energy Extension Service
250 Capitol Hill Building
Nashville, TN 37219

Texas

Texas A&M University
College Station TX 77849

Washington

Energy Extension Service
Washington State University
Pullman, WA 99164

Wisconsin

University of Wisconsin Extension
Energy Programs Coordinator
437 North Lake Street
Madison, WI 53705

Wyoming

Energy Extension Service
University Station
Box 3295
Laramie, WY 82071

5. Alternative Energy Organizations (Private)

Arizona

Advance Solar Energy Technology
1609 West Windrose
Phoenix, AZ 85029

California

Citizen's Energy Conservation
and Solar Development, Inc.
PO Box 49173
Los Angeles, CA 90049

Portola Institute
558 Santa Cruz Avenue
Menlo Park, CA 94025

Farallones Institute
15290 Coleman Valley Road
Occidental, CA 95465

Wind Energy Society of America
1700 East Walnut Street
Pasadena, CA 91106

Friends of the Earth —
Energy Department
124 Spear Street
San Francisco, CA 94111

Solar Energy Society of America
Box 4264
Torrance, CA 90510

Volunteers in Asia, Inc.:
Appropriate Technology Project
PO Box 4543
Stanford, CA 94305

Colorado

Roaring Fork Resource Center
PO Box 9950
Aspen, CO 81611

Environmental Action
2239 East Colfax Avenue
Denver, CO 80206

Alternative Energy Institute
PO Box 3100
Estes Park, CO 80517

Connecticut

Center for the Environment and Man, Inc.
275 Windsor Street
Hartford, CT 06120

District of Columbia

Solar Lobby

American Solar Energy Association
Suite 931
1346 Connecticut Avenue
Washington, DC 20036

Americans for Energy Independence
1629 K Street, NW
Suite 1201
Washington, DC 20006

Appropriate Technology International
1709 N Street, NW
Washington, DC 20036

Bio-Energy Council
Suite 204
1337 Connecticut Avenue, NW
Washington, DC 20036

Center for Science in the Public Interest
1779 Church St. NW
Washington, DC 20007

Citizens' Energy Project
1413 K Street, NW
8th Floor
Washington, DC 20005

CONCERN
2233 Wisconsin Avenue, NW
Washington, DC 20007

Environmental Law Institute
1346 Connecticut Avenue NW
Washington, DC 20036

Environmentalists for Full Employment
1785 Massachusetts Avenue NW
Washington DC 20036

Institute for Local Self-Reliance
1717 18th Street NW
Washington, DC 20009

Public Interest Research Group
1346 Connecticut Avenue, NW
Washington, DC 20036

Solar Energy Institute of America
PO Box 9852
Washington, DC 20005

Solar Energy Research and Education Foundation
1001 Connecticut Avenue, NW
Suite 800
Washington, DC 20036

Solar Energy Research and Information Center, Inc.
1001 Connecticut Avenue, NW
Washington, DC 20036

Solar Lobby
1028 Connecticut Avenue, NW
Washington, DC 20036

The Solar Solution, Inc.
6802 Walker Mill Road, SE
Washington, DC 20027

World Future Society
PO Box 30369
Bethesda Branch,
Washington, DC 20014

Illinois

Institute of Environmental Sciences
940 East NW Highway
MT. Prospect, IL 60056

Acorn
Governor's State University
Park Forest South, IL 60466

Indiana

American Wind Energy Association
54468 CR 31
Bristol, IN 46507

Environic Foundation International, Inc.
PO Box 88
Notre Dame, IN 46556

Maryland

Volunteers in Technical Assistance
3706 Rhode Island Avenue
Mount Rainier, MD 20822

Massachusetts

New Alchemy Institute
PO Box 47
Woods Hole, MA 02543

Minnesota

Alternative Sources of Energy
Route 2, Box 90A
Milaca, MN 56353

American Underground-Space Association
Department of Civil and Mineral Engineering
University of Minnesota
Minneapolis, MN 55455

Montana

Alternative Energy Resource Organization
435 Stapleton Building
Billings, MT 59101

National Center for Appropriate Technology
PO Box 3838
Butte, MT 59701

Nebraska

Midwest Energy Alternative
PO Box 83202
Lincoln, NE 68502

New Hampshire

Solar Data
13 Evergreen Road
Hampton, NH 03842

Total Environmental Action, Inc.
Church Hill
Harrisville, NH 03450

New Mexico

Alternative Energy Associations of America
PO Box 26507
Albuquerque, NM 87125

New York

Consumer Action Now — Council on Environmental Alternatives
355 Lexington Avenue
New York, NY 10017

Energy Task Force, Inc.
156 Fifth Avenue
New York, NY 10010

Environment Information Center, Inc.
124 East 39th Street
New York, NY 10016

Center for Energy Policy
and Research
New York Institute of Technology
Old Westbury, NY 11568

Oregon

The Delphian Foundation
Route 2, Box 195
Sheridan, OR 97378

Virginia

National Recreation and
Park Association
1601 North Kent Street
Arlington, VA 22209

Vermont

Wood Energy Institute
PO Box 1, Fiddlers Green
Waitsfield, VT 05673

Washington

2332 East Madison
Seattle, WA 98122

6. Miscellaneous Energy-Related Services by State

Arizona

Solar Energy Collection, Arizona State University Library
Tempe, Arizona 85281
(602) 965 7608

Areas of Interest: Solar thermal utilization: photovoltaic applications.

Holdings: Archives of the International Solar Energy Society research collection, technical papers: research reports, patents.

Information Services: Answers brief inquiries, library collection available for public use (free). Provides computerized information searches. *Solar Energy Index* (microfiche index to their holdings).

Colorado

Environmental Action Reprint
Reprint Service — (EARS)
2239 East Colfax
Denver, Colorado 80206
(303) 321 1645
Operates a solar bookstore: send for a free catalog.

Areas of Interest: Solar and alternative energy information.

Typical Publications: *Solar Energy Handbook; An Attached Solar Greenhouse; Power from the Wind; Solar Science Projects.*

Solar Energy Applications Laboratory, Colorado State University
Fort Collins, Colorado 80523
(303) 491 8618

Areas of Interest: Solar energy applications.

Holdings: Access to Colorado State University Library Collection

Publications: Over 30 technical reports including *Design and Construction of a Residential Solar Heating and Cooling System.* Send for free publication list.

Information Services: Will answer inquiries from public and in-depth technical questions from scientists/engineers (free). Write for brochure describing laboratory activities.

Delaware

Energy Information Services,
Institute of Energy Conversion
University of Delaware
1 Pike Creek Center
Wilmington, Delaware 19711
(302) 994-0915 or 994-0916

Areas of Interest: Solar energy conversion in relation to distributed conversion systems for heating/cooling and for electrification using photovoltaic solar cells.

Publications: Reports of the institute's activities, pamphlets.

Information Services: Answers inquiries or makes referrals to other sources of information, distributes publications (free). Provides consulting services and conducts workshops.

District of Columbia

Solar Energy Industries Association
1001 Connecticut Ave., Suite 800
Washington, D.C. 20036
(202) 293 2981

Areas of Interest: Manufacturing, design, and installation of solar systems; heating and cooling; solar energy.

Publications: *SEIA News* (monthly); *Solar Industry Index* (annual); *Solar Engineering* (monthly); *Solar-Industry Advertising Guidelines*

Information Services: Answers inquiries free: makes free referrals: market studies, position papers, analyses.

New York

American Society of Heating, Refrigerating and Air-Conditioning Engineers, Inc.
345 East 47th Street
New York, New York 10017
(212) 644-7500

Areas of Interest: Solar energy utilization.

Holdings: Collection of Engineering Societies Library includes 110,000 books, standards, specifications, and codes.

Publications: *ASHRAE journal*; free list of publications.

Information Services: Engineering Societies Library collection available for on-site use. Answers inquiries from professional engineers (free).

Environmental Action Coalition
235 East 49th Street
New York, New York 10017
Send stamped self-addressed business envelope to above address for more information and prices.

Areas of Interest: Educational packets on all aspects of energy use.

Typical Publications: Educational packets, newsletters, films, children's newsletter *Eco-News*.

7. U.S. Chapters of the International Solar Energy Society Listed by State and/or Region

The American Section chapters of the ISES provide information and participatory opportunity for interested persons. Call or write for membership and price information. ISES relates to all uses and applications of solar energy.

Publications include *Solar Age* (monthly); *Solar Energy* (bimonthly); *Sun World* (quarterly); *ISES News* (international newsletter); annual membership directory; conference proceedings, books, reprints, educational materials.

The headquarters for ISES — American Section is the American Technological University, PO Box 1416, Killeen, Texas 76541 (817-526-1300). The International Solar Energy Society's address is PO Box 52, Parkville, Victoria, Australia 3052.

Alabama

Solar Energy Association
c/o University of Alabama in Huntsville
Johnson Environmental & Energy Studies
PO Box 1247
Huntsville, AL 35807

Arizona

Solar Energy Association
Chairman
Arizona State University
Tempe, AZ 85281

California (Northern California)

Solar Energy Association
Director
Center for Solar Energy Applications
San Jose State University
San Jose, CA 95192

California (Southern California)

Solar Energy Association
202 C Street 11 B
San Diego, CA 92101

Colorado

Solar Energy Association
c/o Solar Energy Program
University of Colorado
1100 14th Street
Denver, CO 80202

Georgia

Solar Energy Association
Campus Box 32743
Georgia Institute of Technology
Atlanta, GA 30332

Illinois (Northern Illinois)

Solar Energy Association
PO Box 352
Argonne National Laboratory
Argonne, IL 60439

Michigan

Solar Energy Association
201 East Liberty Street
Ann Arbor, MI 48104

Mid-Atlantic

Solar Energy Association
Department of Architecture
Graduate School of Fine Arts
University of Pennsylvania
Philadelphia, PA 19104

New England

Solar Energy Association
PO Box 541
Brattleboro, VT 05301
(802) 254-2386

New Mexico

Solar Energy Association
PO Box 2004
Santa Fe, NM 87501

New York (Eastern New York)

Solar Energy Society
PO Box 5181
Albany, NY 12205
(518) 457-4861

New York (Metropolitan New York)

Solar Energy Association
United Engineering Center
345 East 47th Street
New York, NY 10017

Ohio

Solar Energy Association
12931 West Geauga Trail
Chesterland, OH 44026
(216) 729-9350

Texas

Solar Energy Society
Texas A&M University
Mechanical Engineering Department
College Station, TX 77843

Virginia

Solar Energy Association
PO Box 12442
Richmond, VA 23241

Public Law 95-6187
8. The Energy Tax Act of 1978

The Energy Tax Act of 1978 provides tax credits for certain improvements made by homeowners between April 20, 1977, the date of President Carter's first energy message, and 1985. A tax credit is subtracted from the amount due the government after the taxes are computed.

Homeowners who install solar, wind or geothermal energy equipment are eligible for a maximum credit of $2,200. For up to $2,000 in expenses, a homeowner could take a 30 percent credit, plus a 20 percent credit on expenses between $2,000 and $10,000.

The act also provides a maximum tax credit of 15 percent on the first $2,000 spent by homeowners on insulation, storm windows, caulking, weather stripping and other specified energy-saving measures — a maximum of $300.

The act also gives businesses a break — a nonrefundable 10 percent credit effective October 1, 1978. This credit is available for boilers, burners for combustors other than boilers, geothermal power equipment (not including turbines or generators), equipment for producing synthetic gas, pollution control equipment required by regulation, solar or wind energy equipment and equipment for heat recovery, recycling, shale oil production and geopressurized methane production.

9. Books

Alternative Sources of Energy: Practical Technology and Philosophy for Decentralized Society. Alternative Sources of Energy, Box 90A, Milaca, NY (1974)

Sometimes it is hard to relate philosophy to the practicalities of life. The author of this book combines both.

Architectural Graphics. Frank Ching. Van Nostrand Reinhold, NY (1975)

Communication of design ideas via paper to the builders.

Building Construction Illustrated. Frank Ching. Van Nostrand Reinhold, NY (1975)

Step-by-step guidance on each phase of construction of a residential building, cost-estimating, and construction sequencing. Tables also show you how to calculate building loads.

The Coming Age of Solar Energy. Daniel Halacy, Jr. Harper and Row, NY (1964)

An important introduction to the use of solar energy and how it will affect us.

Design with Climate. Victor Olgyay. Princeton University Press, Princeton, NJ (1963)

There is more to living with alternative energy than putting a collector on your roof. Olgyay discusses designing homes conservatively to suit your particular climate and to take full advantage of building orientation and natural shelter.

Designing and Building a Solar House. Donald Watson. Gardenway Publishing, Charlotte, VT 05445 (1977)

Detailed explanations of solar designs and systems with photographs and drawings.

Direct Use of the Sun's Energy. F. Daniels. Ballantine Paperback, NY (1974)

A very useful source book of information.

Earth Sheltered Housing Design: Guidelines, Examples, and References. The Underground Space Center, University of Minnesota, Minneapolis, MN

If you plan to build a solar house, consider doing an *underground* passive solar house. The advantages are many. This book provides data on all aspects of underground housing, including a section discussing various houses already built.

Energy Essays. Malcolm Wells. Edmund Scientific Co., 150 Edscorp Building, Barrington, NJ 08007

A collection of essays on the many aspects of energy use and conservation. These essays are especially important if you wonder what all the fuss over energy is about.

Energy for Survival. Wilson Clark. Anchor Books, Garden City, NY 11530 (1974)

Philosophical approach to the need for conservation and development of alternative energy. If you plan to use alternative energy (and practice conservation), things will go much easier if you understand why it is necessary.

Energy Primer. Portola Institute, Menlo Park, CA 94025 (1975)

Clear, comprehensive information on many small energy systems, including solar.

The Food and Heat Producing Solar Greenhouse. Rick Fisher and Bill Yanda. John Muir Publications, Berkeley, CA 94710 (1977)

The greenhouse can be an important heating element in your solar house. Full details on constructing and using an attached greenhouse. Illustrations, lists, and construction diagrams are a great help.

Handmade Houses: A Guide to the Woodbutcher's. Art Boericke and Barry Shapiro. Scrimshaw Press, San Francisco, CA (1973)

A collection of owner-designed and -built homes that show creativity and ingenuity in layout and use of materials.

Homeowner's Guide to Solar Heating. Sunset Books, Lane Publishing, Menlo Park, CA 94025 (1978)

An excellent guide for the homeowner with extensive and valuable information on specific systems and many individual solar homes.

How Things Don't Work. Victor Papanek and James Hennessey. Pantheon, NY (1977)

A seriously funny book about the technology we live with and how it can be improved. Seriously consider these comments before building.

Living with the Sun: A Guide for Solar Home Heating. William Davis. Solar Energy Services, PO Box 2166, Hendersonville, NC 28739

A very readable collection of information including a survey of existing collector systems, how to design a solar heating system, and appendices full of the data necessary for designing a system.

Natural Solar Architecture: A Passive Approach. David Wright, Van Nostrand Reinhold, NY (1978)

David Wright is an advocate of passive solar houses and has designed and built a number of them. He currently lives in an underground passively heated home. His ideas are worth studying.

New Inventions in Low-Cost Solar Heating: 100 Daring Schemes Tried and Untried. William A. Shurcliff, Brick House Publishing Co., Andover, MA 01810 (1979)

A collection of imaginative new schemes for solar heating. Dr. Shurcliff, inventor and Harvard professor, surveys radical new inventions by others and describes many new schemes of his own. His emphasis is on simple, low-cost systems: new passive systems; active systems that are cheap; new ideas on storage systems, hot-water systems and greenhouses. A book for everyone interested in solar.

Shelter I & II. Shelter Publications, P.O. Box 279, Bolinas, CA 94924 (1973 & 1978)

Shelter surveys housing around the world, past and present. The articles and tables provide good sound data and references. The pictures, plans and designs are excellent idea resources for creative shelter design. A third edition is in planning.

Solar Energy: Fundamentals in Building Design. Bruce Anderson. TEA Inc., Church Hill, Harrisville, NH 03450 (1977)

This is an excellent source book of information on solar energy and its applications in building design. The material is pertinent for both beginners and architects. It is wise for anyone planning to use alternative energy to have an understanding of what goes into the design and construction of a solar-heated house.

Small Is Beautiful. E. F. Schumacher. Harper & Row, NY (1973)

The philosopher of living in a low-energy, appropriate-technology society. Schumacher makes a great deal of sense and should be read by everyone in an energy-conserving environment.

Solar Energy Thermal Processes. John A. Duffie and William

Beckman. John Wiley and Sons, Inc. NY (1973)

This is an engineers' book, but if you can wade through the material, it is an excellent reference resource.

Solar Guide: Inexpensive, Economical Solar Heating Systems for Homes. National Technical Information Service, 5285 Port Royal Rd., Springfield, VA 22150 (1977)

Provides complete directions for adding solar heating to an existing forced-air system at a cost of $2,000 (in 1977 dollar values).

Solar Heated Buildings of North America: 120 Outstanding Examples. William A. Shurcliff. Brick House Publishing Co., Andover, MA 01810 (1978)

Covers private residences in all price ranges as well as some commercial and government buildings. Each of the 120 entries includes accurate and concise descriptions of the building, its solar collection system, and the storage system. An up-to-date view of solar heat in America.

Solar Heating and Cooling Systems. Jan L. Kreith and Frank Kreider. Hemisphere Publishing Corp., Washington, DC (1975)

A book for the technician, engineer, and advanced do-it-yourselfer. This book contains a lot of useful design and mechanical information.

The Solar Home Book. Bruce Anderson. Brick House Publishing Co., Andover, MA 01810 (1976)

An invaluable reference for amateur and professional with clear, inclusive information on solar heating and cooling, solar references, products and supplies.

Solar House Heating and Air-Conditioning Systems: Comparisons and Limitations. Harry E. Thomason and Harry J. Thomason. Edmund Scientific Co., Barrington, NJ (1974)

The Thomasons were pioneers in the early modern days of using solar energy for heating and cooling. Their experience is invaluable.

Underground Designs. Malcolm Wells. PO Box 1149, Brewster, MA 02631

If you are interested in using solar energy for heating and cooking, underground is the way to go. This book contains designs both practical and not so practical that demonstrate the beauty to be achieved in a Hobbit-hole. Underground dwellings make a lot of sense.

The Use of Earth-Covered Buildings. Proceedings and Notes of a Conference Held in Ft. Worth, TX, July 9-12, 1975. Supt. of Documents, U.S. Government Printing Office, Washington, DC 20402

This collection of papers is important material for anyone planning to build a solar (or any other kind) of underground house.

Your Home's Solar Potential: The Do-It-Yourself Solar Survey. Irwin Spetgang and Malcolm Wells. Edmund Scientific Co., Barrington, NJ (1976)

Tells how to take a look at your house and determine its potential for solar heating devices. More importantly, it provides the means for analyzing your energy consumption, shows how to conserve and cut back on energy use, and gives costs in 1976 dollars.

10. Solar Energy Periodicals

Most of these solar-energy publications are available by subscription. Write to the addresses listed for more information.

Alternative Sources of Energy
Route 2, Box 90A
Milaca, MN 56353
Monthly. Excellent information on practical, do-it-yourself, uses of alternative energy.

Colorado Solar Energy Association Newsletter
Colorado Solar Energy Assoc.
c/o Solar Energy Program
University of Colorado at Denver
1100 14th Street
Denver, CO 80202
Monthly. Solar projects & events in Colorado. Excellent source of material for community & low-technology projects.

Mother Earth News
PO Box 70
Hendersonville, NC 28739
Bi-monthly. Timely articles on all alternatives: economics, housing, energy, transportation, gardening/farming. Best source of books for alternatives.

RAIN: Journal of Appropriate Technology
2270 Northwest Irving St.
Portland, OR 97210
Monthly. Emphasizes small-scale home-built and community projects.

Solar Age
Solar Vision, Inc.
Church Hill
Harrisville, NH 03450
Monthly. Informative articles on all aspects of the use of solar

energy. New products guide, business analysis.

Solar Energy
ISES Publication
Pergamon Press
Maxwell House, Fairview Park
Elmsford, NY 10523
Bi-monthly. Free to members of ISES. Technical articles on solar applications around the world.

Solar Energy Digest
PO Box 17776
San Diego, CA 92110
Monthly. Nontechnical coverage of new developments in solar energy. Excellent for resources & analysis of products & projects.

Solar Engineering
Suite 880,
8435 N. Stemmons Freeway
Dallas, TX
Monthly. Publication of Solar Energy Industries Assn. Articles on solar applications, products.

Solar Heating and Cooling
Gordon Publications, Inc.
20 Community Place
Morristown, NJ 07960
Bi-monthly journal. Covers basic practical problems specifically in solar heating and cooling.

Solar Utilization News
Alternative Energy Institute
PO Box 3100
Estes Park, CO 80517
Bi-monthly. National solar news articles on projects, events, problems, bibliographies on selected subjects. Some technical articles.

Southwest Bulletin
New Mexico Solar
Energy Association
PO Box 2004
Santa Fe, NM 87501
Monthly. Excellent articles on low-technical solar projects in the southwest.

Sun Times
Alternative Energy Resources
Organization
435 Stapleton Building
Billings, MT 59101
Monthly. Newsletter, includes how-to and some technical articles.

Sun Up
An "Energy News Digest"
J. Harrington and Associates
58913 Carmelita Circle
Yucca Valley, CA 92284
Monthly. News on national solar events, projects, products; primarily for solar trade.

Sunworld
ISES
PO Box 26
Highett, Victoria 3190
Australia
$13.20 per year.
Subscription through Pergamon Press, Inc., Maxwell House, Fairview Park, Elmsford, NY 10523
Quarterly. World-wide development & use of solar; technical & non-technical, practical & theoretical coverage of low-technology ideas, projects.

11. Additional Resource Material for the Solar Consumer

Books, reports, and papers on solar energy are also available from the private concerns listed here. Send for their catalogs.

Arthur D. Little, Inc.
Acorn Park
Cambridge, MA 02140

Brick House Publishing Co.
3 Main St.
Andover, MA 01810

Bruce Research Institute
McGill University
Anne de Bellevue
Montreal, PQ, Canada

Edmund Scientific Co.
150 Edscorp Building
Barrington, NJ 08007

Environmental Action Reprint
Service
2239 East Colfax Avenue
Denver, CO 80206

Garden Way Publishing Co.
Charlotte, VT 05445

New Alchemy Institute
Box 432
Woods Hole, MA 02543

Portola Institute
Menlo Park, CA 94025

Rodale Books
Emmaus, PA 18049

TAB Books
Blue Ridge Summit, PA 17214

TEA, Inc.
Church Hill
Harrisville, NH 03450

Zomeworks, Inc.
Box 712
Albuquerque, NM

12. Glossary

Absorber plate — A metal plate in a collector that absorbs the sun's radiation.

Active system — A system that requires external mechanical power to move the heat that has been collected. Active systems are not integrated into the heating system of a dwelling.

Air-type system — A collection system using air as the medium for transferring heat.

Angle of inclination — See Tilt.

Auxiliary or backup system — A heating system used to supplement a solar system when the latter cannot supply the full heating requirement.

Blower or fan — A device used to create a current of air.

BTU or British thermal unit — A unit of energy equal to the amount of heat needed to raise the temperature of one pound of water 1°F.

Closed system — A system in which there is no direct connection between the fluid circulating in the collector and the fluids used in the household.

Coating — See Selective surface.

Coils — A series of connected pipes arranged in a spiral or circular shape.

Collector — Any device used to collect solar radiation.

Concentrating collector — A collector which uses lenses or reflecting surfaces to focus and intensify solar radiation.

Conduction — A process transferring heat through the collection, transportation, storage or use of solar energy.

Controls — Devices used to regulate the collection, transportation, storage or use of solar energy.

Convection — The process by which a liquid or gas transfers heat from one place to another by circulation.

Degree days or D.D. — The number of degrees by which the daily mean outside air temperature is below 65°F. during a given heating season.

Diffuse radiation — Radiation that is scattered in solid or liquid particles in the atmosphere.

Direct conversion — Converting sunlight directly into electric power.

Direct radiation — Radiation that comes directly from the sun without deflection or scattering in the atmosphere.

Distribution system — An arrangement to disperse solar heat for use within a building.

Distillation of water — The controlled evaporation of water from a combination of water and other substances, using the condensation of the resulting water vapor to provide purified water.

Double-glazed — Having two layers of glass in windows, doors or other openings, usually with a sealed air space between them.

Drying, solar — An agricultural or industrial process by which moisture is removed from a substance through use of solar energy.

Duct — A tube, pipe, or channel for carrying liquids or gases.

Efficiency of collector — The ratio of the amount of heat extracted by the collector to the amount of heat falling on the surface of the collector.

Filter — A device for straining substances from liquid or gas by passing it through special materials.

Flashing — A sheet of metal or other material used for protection against water seepage.

Flat-plate collector — A flat surface for the collection of solar heat.

Header — A pipe, running across the top or bottom of a collector, which gathers liquid for transferring heat.

Heat exchanger — A device used to transfer heat from one medium to another.

Heat pump — A mechanical device used to transfer heat from one material to another by cooling the first and warming the second.

Heat transfer fluid — The liquid or gas that transfers heat from the collector to the point of storage or use.

Incidence — See Tilt.

Infiltration — The amount of outside air entering a building through windows, doors, cracks and other openings.

Insolation — A technical term for the amount of radiation falling on a surface, including direct, diffused and reflected radiation.

Insulation — A nonconducting material used to reduce the flow of heat through walls, ceilings, floors, etc.

Langley — A unit for measuring solar radiation, equal to one calorie of radiation per square centimeter.

Life-cycle costing — A way of measuring long-term costs by including maintenance costs, repairs, and energy consumption.

Liquid-type collector — A collector that uses a liquid as the medium for transferring heat.

Microclimate — Data on insolation and weather in a very small geographical area, taking into account altitude, cloudiness, the presence of bodies of water, etc.

Nocturnal radiation — The loss of heat at night from a warm building to the colder outside air and other surroundings.

Open system — A system in which the same fluid is used throughout.

Passive system — A system that uses the design and mass of the building as a heating system, without the use of mechanical devices for moving and distributing heat.

Photosynthesis — A natural process which captures and stores energy by the chemical action of chlorophyll in plants and algae.

Photovoltaic cells — A solar system that uses a chemical reaction to convert solar energy directly into electricity.

Pyranometer — An instrument for measuring total insolation both direct and diffused.

Pyrheliometer — An instrument that can measure direct insolation and reflectance.

Pump — An apparatus used for forcing a liquid or gas into, or out of, something. (See also Heat pump)

Radiation — Energy given off by the sun in the form of heat and light.

Reflectance or reflectivity — The ratio of light falling on a surface to the amount which is not absorbed.

Reflected radiation — The radiation given off by buildings, water, or other surrounding objects which reflect the sun's rays.

Register — An opening with a device to regulate the amount of heat or air flowing through it.

Resistance or R-value — A measure of thermal resistance, or the ability of a material to retard the flow of heat.

Retrofit — The installation of a solar heating or cooling system in an existing structure.

Selective surface or coating — An optical coating used on collectors to increase absorption and decrease reradiation.

Sensor — A device that reacts to changes in temperature and can relay this information to a thermostat.

Solar domestic hot water heating, or solar domestic hot water system, or S.H.W. — A system for heating hot water by using solar energy.

Solar house — A house that gains a significant portion of its heat from the sun.

Solar heating system — A system using the sun for heating the inside of a building. Usually includes: space heating and cooling and often hot water heating or some combination of these. Usually contains collectors, storage facilities, controls and instrumentation, and a distribution system consisting of pipes, vents, ducts and fans.

Storage — A space or system designed to receive and preserve heat for future use.

Storage medium — The material, liquid, solid or both, used for storing collected solar heat.

Sun rights — Legal rights guaranteeing continuous, unobstructed access to the sun to solar builders and dwellers.

Switch — A device for turning an electric current on or off.

Thermal — Of or pertaining to heat.

Thermal wall — A masonry wall used to store heat from the sun.

Tilt (Also, Incidence or Angle of inclination) — The angle at which a flat-plate collector is tilted for obtaining maximum exposure to the sun and maximum collection of solar radiation.

Triple-glazing — A triple thickness of glass (in windows, doors, or other openings), ordinarily with sealed air spaces between the sheets.

U-value — A measure of thermal conductance based on the amount of heat lost through a square foot of wall.

Valve — A movable part which controls the flow of a liquid or gas.

Vapor barrier — Material used to retard the movement of water vapor into walls and to prevent condensation.

Variance — A procedure used to allow exceptions to existing building codes due to unusual circumstances.